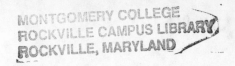
STRUCTURED ANALYSIS

AND

SYSTEM SPECIFICATION

by

Tom DeMarco

Foreword by

P.J. Plauger

YOURDON inc.
1133 Avenue of the Americas
New York, New York 10036

First Printing, March 1978

Second Printing, June 1978

Revised, December 1978

ISBN: 0-917072-07-3

DEDICATION

To G. W. F.,
with my respect and admiration and affection.

CONTENTS

PART 3: DATA DICTIONARY

PART 4: PROCESS SPECIFICATION

PART 5: SYSTEM MODELING

PART 6: STRUCTURED ANALYSIS FOR A FUTURE SYSTEM

ACKNOWLEDGMENTS

In the Afterword to his epic tale on the French Revolution, *Ninety Three,* Victor Hugo describes his writing method. He wrote standing at upright desk in pen and ink on oversize sheets of white paper. When he had polished and completed his manuscript, he sent it to the printer. No editors, no proofreaders, no layout artists. He seemed to need no one to correct his spelling or untangle his syntax. But then, he was Victor Hugo. Tom DeMarco needed help.

My most important helper has been Toni Nash. Her great enthusiasm and energy have kept the whole project moving. She translated my manuscript into English from whatever language it was in before. She acted as editor for the effort, and, as well, as its technical consultant. I can't think of anyone who could have done the two jobs better: to clean up the presentation and punctuation on one page, and then to check balancing and data conservation of Data Flow Diagrams on the next.

My thanks to Ed Yourdon, Jerry Walker, Bill Plauger, and Lois Rose (all of Yourdon inc.) and to Mark Bade and Sally Proesel (of Deltak Inc.) for wading through the early manuscripts and offering comments and suggestions. Thanks, too, to Barbara Smyth for her cover design and to Margot Schwaab for her help in assembling the final work.

Finally, I owe a large debt of thanks to past students of my analysis seminars. They have been an unfailing source of insight, constructive criticism, and useful ideas.

New York
February 1978

Tom DeMarco

FOREWORD

It is a pure pleasure to watch the emergence of a new discipline. "Structured Analysis" has progressed from being a concatenation of buzzwords, to a collection of homilies, to an orderly approach that commands the respect of the most battle-scarred senior analysts. I know of no more critical audience, nor one so in need of technical assistance, than today's systems analysts; Structured Analysis addresses their needs laudably.

When Tom DeMarco joined forces with YOURDON inc. almost two years ago, he brought with him a wealth of EDP experience, a resume that would make your mouth water, and the prettiest collection of neckties I have ever admired. He has proved to be a seminar leader of the first rank thanks to an uncluttered delivery, astonishing candor, and — that most important gift of all — an eagerness to learn from his students. In this environment, he has transmuted the YOURDON approach to analysis into a discipline of compelling elegance.

What I like most about this book is how well it teaches the construction and evaluation of Data Flow Diagrams. Larry Constantine encouraged the use of such graphic aids over a dozen years ago as a way to analyze a restricted class of systems known as transform centers. Transform Analysis, however, seemed to get lost among the myriad innovations of Structured Design. It is only with 20-20 hindsight that we can see that the transform center is but the simplest non-trivial Data Flow Diagram, and that an understanding of data flow is vital to the success of *any* system design.

The pragmatics of building a Data Dictionary are equally important, for just as the data flow forms the backbone of a system, so too does the data *structure* rigorously shape the innards of each module. Or at least it should. Michael Jackson has taught us how to progress in an orderly fashion from a clear understanding of the data being processed by a module to a well structured program that expressed the process. Structured Analysis provides the context in which that more localized approach can thrive.

DeMarco has also delivered up a most readable book, no surprise if you have ever heard the man speak, but a violation of Industry Standards nonetheless. When I was asked to write this foreword, I set aside a few hours to skim the manuscript. I ended up staying up half the night reading it in detail, alternately nodding my head at telling points and chuckling at the aptness of anecdotes. He is candid when ignorant, and convincing when he himself is convinced. I look forward to rereading this thing when the ink dries.

My greatest concern is that the profundity of what this book is saying will be lost because it comes out so smoothly. I want to look over each reader's shoulder so that I can repeatedly stab my finger at selected pages and say, "There! Read that again! Think what it implies!" Fortunately for the peace and quiet of offices and armchairs, I can only do that on this page. Besides, Structured Analysis shouldn't be so nerve wracking. Let me say simply: Listen to the man, think about what he has to say, and for heaven's sake enjoy.

New York
February 1978   P. J. Plauger

PART 1

BASIC CONCEPTS

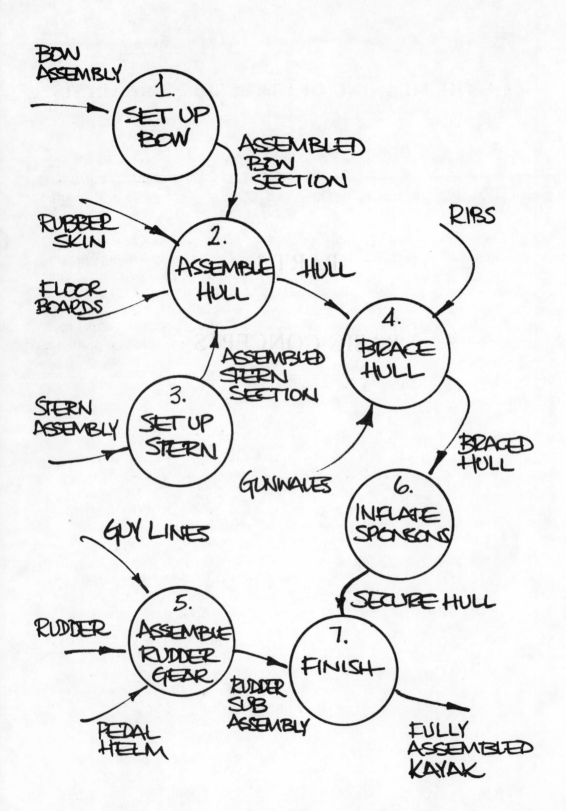

Figure 1

1 THE MEANING OF STRUCTURED ANALYSIS

Let's get right to the point. This book is about Structured Analysis, and Structured Analysis is primarily concerned with a new kind of Functional Specification, the Structured Specification. Fig. 1 shows part of a Structured Specification.

The example I have chosen is a system of sorts, but obviously not a computer system. It is, in fact, a manual assembly procedure. Procedures like the one described in Fig. 1 are usually documented by a text narrative. Such descriptions have many of the characteristics of the classical Functional Specifications that we analysts have been producing for the last 20 years. (The Functional Specification describes *automated* procedures — that is the main difference between the two.) Take a look at a portion of the text that prompted me to draw Fig. 1.

Assembly Instructions for KLEPPER Folding Boats

1. Lay out hull in grass (or on carpet). Select a clean, level spot.

2. Take folded bow section (with red dot), lay it in grass, unfold 4 hinged gunwale boards. Kneel down, spread structure lightly with left hand near bow, place right hand on pullplate at *bottom* of hinged rib, and set up rib gently by pulling towards center of boat. Deckbar has a tongue-like fitting underneath which will connect with fitting on top of rib if you lift deckbar lightly, guide tongue to rib, press down on deckbar near bow to lock securely. Now lift whole bowsection using both arms wraparound style (to keep gunwales from flopping down) and slide into front of hull. Center seam of blue deck should rest on top of deckbar.

3. Take folded stern section (blue dot, 4 "horseshoes" attached), unfold 4 gunwales, set up rib by pulling on pullplate at *bottom* of rib. Deckbar locks to top of rib *from the side* by slipping a snaplock over a tongue attached to top of rib . . .

    And so forth.

The differences are fairly evident: The text plunges immediately into the details of the early assembly steps, while the structured variant tries to present the big picture first, with the intention of working smoothly from abstract to detailed. The Structured Specification is graphic and the text is not. The old-fashioned approach is one-dimensional (written narrative is always one-dimensional), and the structured variant is multidimensional. There are other differences as well; we'll get to those later. My intention here is only to give you an initial glimpse at a Structured Specification.

Now let's go back and define some terms.

1.1 What is analysis?

Analysis is the study of a problem, prior to taking some action. In the specific domain of computer systems development, analysis refers to the study of some business area or application, usually leading to the specification of a new system. The action we're going to be taking later on is the implementation of that system.

The most important product of systems analysis — of the analysis phase of the life cycle — is the specification document. Different organizations have different terms for this document: Functional Specification, External Specification, Design Specification, Memo of Rationale, Requirements Document. In order to avoid the slightly different connotations that these names carry, I would like to introduce a new term here: the Target Document. The Target Document establishes the goals for the rest of the project. It says what the project will have to deliver in order to be considered a success. The Target Document is the principal product of analysis.

Successful completion of the analysis phase involves all of the following:

1.   selecting an optimal target

2.   producing detailed documentation of that target in such a manner that subsequent implementation can be evaluated to see whether or not the target has been attained

3.   producing accurate predictions of the important parameters associated with the target, including costs, benefits, schedules, and performance characteristics

4.   obtaining concurrence on each of the items above from each of the affected parties

In carrying out this work, the analyst undertakes an incredibly large and diverse set of tasks. At the very minimum, analysts are responsible for: user liaison, specification, cost-benefit study, feasibility analysis, and estimating. We'll cover each of these in turn, but first an observation about some characteristics that are common to all the analyst's activities.

1.1.1 Characteristics of Analysis

Most of us come to analysis by way of the implementation disciplines — design, programming, and debugging. The reason for this is largely historical. In the past, the business areas being automated were the simpler ones, and the users were rather unsophisticated; it was more realistic to train computer people to understand the application than to train users to understand EDP technology. As we come to automate more and more complex areas, and as our users (as a result of prevalent computer training at the high school and college level) come to be more literate in automation technologies, this trend is reversing.

But for the moment, I'm sure you'll agree with me that most computer systems analysts are first of all computer people. That being the case, consider this observation: Whatever analysis is, it certainly is not very similar to the work of designing, programming, and debugging computer systems. Those kinds of activities have the following characteristics:

- The work is reasonably straightforward. Software sciences are relatively new and therefore not as highly specialized as more developed fields like medicine and physics.

- The interpersonal relationships are not very complicated nor are there very many of them. I consider the business of building computer systems and getting them to run a rather friendly activity, known for easy relationships.

- The work is very definite. A piece of code, for instance, is either right or wrong. When it's wrong, it lets you know in no uncertain terms by kicking and screaming and holding its breath, acting in obviously abnormal ways.

- The work is satisfying. A positive glow emanates from the programmer who has just found and routed out a bug. A friend of mine who is a doctor told me, after observing programmers in the debugging phase of a project, that most of them seemed "high as kites" much of the time. I think he was talking about the obvious satisfaction programmers take in their work.

The implementation disciplines are straightforward, friendly, definite, and satisfying. Analysis is none of these things:

- It certainly isn't easy. Negotiating a complex Target Document with a whole community of heterogeneous and conflicting users and getting them all to agree is a gargantuan task. In the largest systems for the most convoluted organizations, the diplomatic skills that the analyst must bring to bear are comparable to the skills of a Kissinger negotiating for peace in the Middle East.

- The interpersonal relationships of analysis, particularly those involving users, are complicated, sometimes even hostile.

- There is nothing definite about analysis. It is not even obvious when the analysis phase is done. For want of better termination criteria, the analysis phase is often considered to be over when the time allocated for it is up!

- Largely because it is so indefinite, analysis is not very satisfying. In the most complicated systems, there are so many compromises to be made that no one is ever completely happy with the result. Frequently, the various parties involved in the negotiation of a Target Document are so rankled by their own concessions, they lose track of what a spectacular feat the analyst has achieved by getting them to agree at all.

So analysis is frustrating, full of complex interpersonal relationships, indefinite, and difficult. In a word, it is fascinating. Once you're hooked, the old easy pleasures of system building are never again enough to satisfy you.

1.1.2 The User Liaison

During the 1960's, our business community saw a rash of conglomerations in which huge corporate monoliths swallowed up smaller companies and tried to digest them. As part of this movement, many centralized computer systems were installed with an aim toward gathering up the reins of management, and thus allowing the conglomerate's directors to run the whole show. If you were an analyst on one of these large Management Information System (MIS) projects, you got to see the user-analyst relationship at its very worst. Users were dead set against their functions being conglomerated, and of course that's just what the MIS systems were trying to do. The philosophy of the 60's was that an adversary relationship between the analyst and the user could be very productive, that analysts could go in, as the representatives of upper management, and force the users to participate and comply.

Of course the record of such projects was absolutely dismal. I know of no conglomerate that made significant gains in centralization through a large Management Information System. The projects were often complete routs. Many conglomerates are now spinning off their acquisitions and finding it rather simple to do so because so little true conglomeration was ever achieved. Due to the experience of the 60's, the term Management Information System, even today, is likely to provoke stifled giggles in a group of computer people.

The lesson of the 60's is that no system is going to succeed without the active and willing participation of its users. Users have to be made aware of how the system will work and how they will make use of it. They have to be sold on the system. Their expertise in the business area must be made a key ingredient to system development. They must be kept aware of progress, and channels must be kept open for them to correct and tune system goals during development. All of this is the responsibility of the analyst. He is the users' teacher, translator, and advisor. This intermediary function is the most essential of all the analyst's activities.

1.1.3 Specification

The analyst is the middleman between the user, who decides what has to be done, and the development team, which does it. He bridges this gap with a Target Document. The business of putting this document together and getting it accepted by all parties is specification. Since the Target Document is the analyst's principal output, specification is the most visible of his activities.

If you visit the Royal Naval Museum at Greenwich, England, you will see the results of some of the world's most successful specification efforts, the admiralty models. Before any ship of the line was constructed, a perfect scale model had to be built and approved. The long hours of detail work were more than repaid by the clear understandings that come from studying and handling the models.

The success of the specification process depends on the product, the Target Document in our case, being able to serve as a model of the new system. To the extent that it helps you visualize the new system, the Target Document is the system model.

1.1.4 Cost-Benefit Analysis

The study of relative cost and benefits of potential systems is the feedback mechanism used by an analyst to select an optimal target. While Structured Analysis does not entail new methods for conduct of this study, it nonetheless has an important effect. An accurate and meaningful system model helps the user and the analyst perfect their vision of the new system and refine their estimates of its costs and benefits.

1.1.5 Feasibility Analysis

It is pointless to specify a system which defies successful implementation. Feasibility analysis refers to the continual testing process the analyst must go through to be sure that the system he is specifying can be implemented within a set of given constraints. Feasibility analysis is more akin to design than to the other analysis phase activities, since it involves building tentative physical models of the system and evaluating them for ease of implementation. Again, Structured Analysis does not prescribe new procedures for this activity. But its modeling tools will have some positive effect.

1.1.6 Estimating

Since analysis deals so heavily with a system which exists only on paper, it involves a large amount of estimating. The analyst is forever being called upon to estimate cost or duration of future activities, CPU load factors, core and disk extents, manpower allocation . . . almost anything. I have never heard of a project's success being credited to the fine estimates an analyst made; but the converse is frequently true — poor estimates often lead to a project's downfall, and in such cases, the analyst usually receives full credit.

Estimating is rather different from the other required analysis skills:

- *Nobody is an expert estimator.* You can't even take a course in estimating, because nobody is willing to set himself up as enough of an authority on the subject to teach it.

- *We don't build our estimating skills, because we don't collect any data about our past results.* At the end of a project we rarely go back and carry out a thorough postmortem to see how the project proceeded. How many times have you seen project performance statistics published and compared to the original estimates? In my experience, this is done only in the very rare instance of a project that finishes precisely on time and on budget. In most cases, the original schedule has long since vanished from the record and will never be seen again.

- *None of this matters as much as it ought to anyway,* since most things we call "estimates" in computer system projects are not estimates at all. When your manager asks you to come up with a schedule showing project completion no later than June 1 and using no more than six people, you're not doing any real estimating. You are simply dividing up the time as best you can among the phases. And he probably didn't estimate either; chances are his dates and manpower loading were derived from budgetary figures, which were themselves based upon nothing more than Wishful Thinking.

All these factors aside, estimating plays a key part in analysis. There are some estimating heuristics that are a by-product of Structured Analysis; these will be discussed in a subsequent chapter. The key word here is *heuristic*. A heuristic is a cheap trick that often works well but makes no guarantee. It is not an algorithm, a process that leads to a guaranteed result.

1.1.7 The Defensive Nature of Analysis

In addition to the analysis phase activities presented above, there are many others; the analyst is often a project utility infielder, called upon to perform any number of odd jobs. As the project wears on, his roles may change.

But the major activities, and the ones that will concern us most in this book, are: user liaison, specification, cost-benefit and feasibility analysis, and estimating.

In setting about these activities, the analyst should be guided by a rule which seems to apply almost universally: *The overriding concern of analysis is not to achieve success, but to avoid failure.* Analysis is essentially a defensive business.

This melancholy observation stems from the fact that the great flaming failures of the past have inevitably been attributable to analysis phase flaws. When a system goes disastrously wrong, it is the analyst's fault. When a system succeeds, the credit must be apportioned among many participants, but failure (at least the most dramatic kind) belongs completely to the analyst. If you think of a system project that was a true rout — years late, or orders of magnitude over budget, or totally unacceptable to the user, or utterly impossible to maintain — it almost certainly was an analysis phase problem that did the system in.

Computer system analysis is like child-rearing; you can do grievous damage, but you cannot ensure success.

My reason for presenting this concept here is to establish the following context for the rest of the book: The principal goal of Structured Analysis is to minimize the probability of critical analysis phase error. The tools of Structured Analysis are defensive means to cope with the most critical risk areas of analysis.

1.2 Problems of analysis

Projects can go wrong at many different points: The fact that we spend so much time, energy, and money on maintenance is an indication of our failures as designers; the fact that we spend so much on debugging is an indictment of our module design and coding and testing methods. But analysis failures fall into an entirely different class. When the analysis goes wrong, we don't just spend more money to come up with a desired result — we spend *much* more money, and often don't come up with any result.

That being the case, you might expect management to be super-conservative about the analysis phase of a project, to invest much more in doing the job correctly and thus avoid whole hosts of headaches downstream. Unfortunately, it is not as simple as that. Analysis is plagued with problems that are not going to be solved simply by throwing money at them. You may have experienced this yourself if you ever participated in a project where too much time was allocated to the analysis phase. What tends to happen in such cases is that work proceeds in a normal fashion until the major products of analysis are completed. In the remaining time, the project team spins its wheels, agonizing over what more it could do to avoid later difficulties. When the time is finally up, the team breathes a great sigh of relief and hurries on to design. Somehow the extra time is just wasted — the main result of slowing down the analysis phase and doing everything with exaggerated care is that you just get terribly bored. Such projects are usually every bit as subject to failures of analysis as others.

I offer this list of the major problems of analysis:

1.   communication problems

2.   the changing nature of computer system requirements

3.   the lack of tools

4.   problems of the Target Document

5.   work allocation problems

6.   politics

Before looking at these problems in more detail, we should note that none of them will be *solved* by Structured Analysis or by any other approach to analysis. The best we can hope for is some better means to grapple with them.

1.2.1 Communication Problems

A long-unsolved problem of choreography is the development of a rigorous notation to describe dance. Merce Cunningham, commenting on past failures to come up with a useful notation, has observed that the motor centers of the brain are separated from the reading and writing centers. This physical separation in the brain causes communication difficulties.

Computer systems analysis is faced with this kind of difficulty. The business of specification is, for the most part, involved in describing procedure. Procedure, like dance, resists description. (It is far easier to demonstrate procedure than to describe it, but that won't do for our purposes.) Structured Analysis attempts to overcome this difficulty through the use of graphics. When you use a picture instead of text to communicate, you switch mental gears. Instead of using one of the brain's serial processors, its reading facility, you use a parallel processor.

All of this is a highfalutin way to present a "lowfalutin" and very old idea: A picture is worth a thousand words. The reason I present it at all is that analysts seem to need some remedial work on this concept. When given a choice (in writing a Target Document, for instance) between a picture and a thousand words, most analysts opt unfailingly for the thousand words.

Communication problems are exacerbated in our case by the lack of a common language between user and analyst. The things we analysts work with — specifications, data format descriptions, flowcharts, code, disk and core maps — are totally inappropriate for most users. The one aspect of the system the user is most comfortable talking about is the set of human procedures that are his interface to the system, typically something we don't get around to discussing in great detail with him until well after analysis, when the user manuals are being written.

Finally, our communication problem is complicated by the fact that what we're describing is usually a system that exists only in our minds. There is no model for it. In our attempts to flesh out a picture of the system, we are inclined to fill in the physical details (CRT screens, report formats, and so forth) much too early.

To sum it up, the factors contributing to the communication problems of analysis are

1.   the natural difficulty of describing procedure

2.   the inappropriateness of our method (narrative text)

3.   the lack of a common language between analyst and user

4.   the lack of any usable early model for the system

1.2.2 The Changing Nature of Requirements

I sometimes think managers are sent to a special school where they are taught to talk about "freezing the specification" at least once a day during the analysis phase. The idea of freezing the specification is a sublime fiction. Changes won't go away and they can't be ignored. If a project lasts two years, you ought to expect as many legitimate changes (occasioned by changes in the way business is done) to occur during the project as would occur in the first two years after cutover. In addition to changes of this kind, an equal number of changes may arise from the user's increased understanding of the system. This type of change results from early, inevitable communication failures, failures which have since been corrected.

When we freeze a Target Document, we try to hold off or ignore change. But the Target Document is only an approximation of the true project target; therefore, by holding off and ignoring change, we are trying to proceed toward a target *without benefit of any feedback.*

There are two reasons why managers want to freeze the Target Document. First, they want to have a stable target to work toward, and second, an enormous amount of effort is involved in updating a specification. The first reason is understandable, but the second is ridiculous. *It is unacceptable to write specifications in such a way that they can't be modified.* Ease of modification has to be a requirement of the Target Document.

This represents a change of ground rules for analysis. In the past, it was expected that the Target Document would be frozen. It was a positive advantage that the document was impossible to change since that helped overcome resistance to the freeze. It was considered normal for an analyst to hold off a change by explaining that implementing it in the Target Document would require retyping every page. I even had one analyst tell me that the system, once built, was going to be highly flexible, so that it would be easier to put the requested change into the system itself rather than to put it into the specification!

Figures collected by GTE, IBM, and TRW over a large sample of system changes, some of them incorporated immediately and others deferred, indicate that the difference in cost can be staggering. It can cost two orders of magnitude more to implement a change after cutover than it would have cost to implement it during the analysis phase. As a rule of thumb, you should count on a 2:1 cost differential to result from deferring change until a subsequent project phase. [1]

My conclusion from all of this is that we must change our methods; we must begin building Target Documents that are highly maintainable. In fact, maintainability of the Target Document is every bit as essential as maintainability of the eventual system.

1.2.3 The Lack of Tools

Analysts work with their wits plus paper and pencil. That's about it. The fact that you are reading this book implies that you are looking for some tools to work with. For the moment, my point is that most analysts don't have any.

As an indication of this, consider your ability to evaluate the products of each project phase. You would have little difficulty evaluating a piece of code: If it were highly readable, well submodularized, well commented, conformed to generally accepted programming practice, had no GOTO's, ALTER's, or other forms of pathology — you would probably be willing to call it a good piece of code. Evaluating a design is more difficult, and you would be somewhat less sure of your judgment. But suppose you were asked to evaluate a Target Document. Far from being able to judge its quality, you would probably be hard pressed to say whether it qualified as a Target Document at all. Our inability to evaluate any but the most incompetent efforts is a sign of the lack of analysis phase tools.

1.2.4 Problems of the Target Document

Obviously the larger the system, the more complex the analysis. There is little we can do to limit the size of a system; there are, however, intelligent and unintelligent ways to deal with size. An intelligent way to deal with size is to *partition*. That is exactly what designers do with a system that is too big to deal with conveniently — they break it down into component pieces (modules). Exactly the same approach is called for in analysis.

---

[1] See Barry Boehm's article, "Software Engineering," published in the *IEEE Transactions on Computers*, December 1976, for a further discussion of this topic.

The main thing we have to partition is the Target Document. We have to stop writing Victorian novel specifications, enormous documents that can only be read from start to finish. Instead, we have to learn to develop dozens or even hundreds of "mini-specifications." And we have to organize them in such a way that the pieces can be dealt with selectively.

Besides its unwieldy size, the classical Target Document is subject to further problems:

- It is excessively redundant.

- It is excessively wordy.

- It is excessively physical.

- It is tedious to read and unbearable to write.

1.2.5 Work Allocation

Adding manpower to an analysis team is even more complicated than beefing up the implementation team. The more successful classical analyses are done by very small teams, often only one person. On rush projects, the analysis phase is sometimes shortchanged since people assume it will take forever, and there is no convenient way to divide it up.

I think it obvious that this, again, is a partitioning problem. Our failure to come up with an early partitioning of the subject matter (system or business area) means that we have no way to divide up the rest of the work.

1.2.6 Politics

Of course, analysis is an intensely political subject. Sometimes the analyst's political situation is complicated by communication failures or inadequacies of his methods. That kind of problem can be dealt with positively — the tools of Structured Analysis, in particular, will help.

But most political problems do not lend themselves to simple solutions. The underlying cause of political difficulty is usually the changing distribution of power and autonomy that accompanies the introduction of a new system. No new analysis procedures are going to make such an impending change less frightening.

Political problems aren't going to go away and they won't be "solved." The most we can hope for is to limit the effect of disruption due to politics. Structured Analysis approaches this objective by making analysis procedures more formal. To the extent that each of the analyst's tasks is clearly (and publicly) defined, and has clearly stated deliverables, the analyst can expect less political impact from them. Users understand the limited nature of his investigations and are less inclined to overreact. The analyst becomes less of a threat.

1.3 The user-analyst relationship

Since Structured Analysis introduces some changes into the user-analyst relationship, I think it is important to begin by examining this relationship in the classical environment. We need to look at the user's role, the analyst's role, and the division of responsibility between them.

1.3.1 What Is a User?

First of all, there is rarely just one user. In fact, the term "user" refers to at least three rather different roles:

- *The hands-on user*, the operator of the system. Taking an on-line banking system as an example, the hands-on users might include tellers and platform officers.

- *The responsible user*, the one who has direct business responsibility for the procedures being automated by the system. In the banking example, this might be the branch manager.

- *The system owner*, usually upper management. In the banking example, this might be the Vice President of Banking Operations.

Sometimes these roles are combined, but most often they involve distinctly different people. When multiple organizations are involved, you can expect the total number of users to be as much as three times the number of organizations.

The analyst must be responsible for communication with *all* of the users. I am continually amazed at how many development teams jeopardize their chances of success by failing to talk to one or more of their users. Often this takes the form of some person or organization being appointed "User Representative." This is done to spare the user the bother of the early system negotiations, and to spare the development team the bother of dealing with users. User Representatives would be fine if they also had authority to accept the system. Usually they do not. When it comes to acceptance, they step aside and let the real user come forward. When this happens, nobody has been spared any bother.

1.3.2 What Is an Analyst?

The analyst is the principal link between the user area and the implementation effort. He has to communicate the requirements to the implementors, and the details of how requirements are being satisfied back to the users. He may participate in the actual determination of what gets done: It is often the analyst who supplies the act of imagination that melds together applications and present-day technology. And, he may participate in the implementation. In doing this, he is assuming the role that an architect takes in guiding the construction of his building.

While the details may vary from one organization to the next, most analysts are required to be

- at ease with EDP concepts
- at ease with concepts particular to the business area
- able to communicate such concepts

1.3.3 Division of Responsibility Between Analyst and User

There is something terribly wrong with a user-analyst relationship in which the user specifies such physical details as hardware vendor, software vendor, programming language, and standards. Equally upsetting is the user who relies upon the analyst to decide how business ought to be conducted. What is the line that separates analyst functions from user functions?

I believe the analyst and the user ought to try to communicate across a "logical-physical" boundary that exists in any computer system project. Logical considerations include answers to the question, *What needs to be accomplished?* These fall naturally into the domain of the user. Physical considerations include answers to the question, *How shall we accomplish these things?* These are in the domain of the analyst.

1.4 What is Structured Analysis?

So far, most of what we have been discussing has been the classical analysis phase, its problems and failings. How is Structured Analysis different? To answer that question, we must consider

- New goals for analysis. While we're changing our methods, what new analysis phase requirements shall we consider?
- Structured tools for analysis. What is available and what can be adapted?

1.4.1 New Goals for Analysis

Looking back over the recognized problems and failings of the analysis phase, I suggest we need to make the following additions to our set of analysis phase goals:

- The products of analysis must be highly maintainable. This applies particularly to the Target Document.

- Problems of size must be dealt with using an effective method of partitioning. The Victorian novel specification is out.

- Graphics have to be used wherever possible.

- We have to differentiate between logical and physical considerations, and allocate responsibility, based on this differentiation, between the analyst and the user.

- We have to build a logical system model so the user can gain familiarity with system characteristics before implementation.

1.4.2 Structured Tools for Analysis

At the very least, we require three types of new analysis phase tools:

- Something to help us partition our requirement and document that partitioning before specification. For this I propose we use a *Data Flow Diagram,* a network of interrelated processes. Data Flow Diagrams are discussed in Chapters 4 through 10.

- Some means of keeping track of and evaluating interfaces without becoming unduly physical. Whatever method we select, it has to be able to deal with an enormous flood of detail — the more we partition, the more interfaces we have to expect. For our interface tool I propose that we adopt a set of *Data Dictionary* conventions, tailored to the analysis phase. Data Dictionary is discussed in Chapters 11 through 14.

- New tools to describe logic and policy, something better than narrative text. For this I propose three possibilities: *Structured English, Decision Tables,* and *Decision Trees.* These topics are discussed in Chapters 15 through 17.

1.4.3 Structured Analysis — A Definition

Now that we have laid all the groundwork, it is easy to give a working definition of Structured Analysis:

Structured Analysis is the use of these tools:

> *Data Flow Diagrams*
> *Data Dictionary*
> *Structured English*
> *Decision Tables*
> *Decision Trees*

to build a new kind of Target Document, the Structured Specification.

Although the building of the Structured Specification is the most important aspect of Structured Analysis, there are some minor extras:

- estimating heuristics

- methods to facilitate the transition from analysis to design

- aids for acceptance test generation

- walkthrough techniques

1.4.4 What Structured Analysis Is Not

Structured Analysis deals mostly with a subset of analysis. There are
many legitimate aspects of analysis to which Structured Analysis does not
directly apply. For the record, I have listed items of this type below:

- cost-benefit analysis

- feasibility analysis

- project management

- performance analysis

- conceptual thinking (Structured Analysis might help you communicate better with the user; but if the user is just plain
  wrong, that might not be of much long-term benefit.)

- equipment selection

- personnel considerations

- politics

My treatment of these subjects is limited to showing how they fit in with the
modified methods of Structured Analysis.

2 CONDUCT OF THE ANALYSIS PHASE

This chapter will look at the development life cycle as it exists in most organizations today, and at the effect that the introduction of Structured Analysis will have on it. The intention of the chapter is to show you a rather different view of Structured Analysis, one which considers its procedures and methods rather than its products.

In all of the following, I shall endeavor to make use of some graphic techniques common to Structured Analysis.

2.1 The classical project life cycle

Fig. 2 represents what I think of as the classical project life cycle, that is, one which involves little or no use of structured technology. This figure would also serve to describe a life cycle that involved some structured coding; but introduction of Structured Design and Structured Analysis, as we shall see, will cause substantial change.

The analysis phase is typically not the first project phase. Usually, as in the figure, it is preceded by some sort of a survey or Feasibility Study, whose purposes are

- to determine whether there is some new way of doing business (new procedures, new devices, new business areas) that justifies the expense of a project

- to document the parameters that would govern such a project

In that the survey tries to define what will be built, and to produce budget and schedule information, it is very like the analysis phase. In fact, the survey in most organizations is a mini-analysis phase. It does all the tasks commonly associated with the analysis phase, but it does them less thoroughly.

The major input to the survey is data gathered from a dialogue with the users. The results of the survey are usually packaged into something called a Feasibility Document, or Technical Requirements Document, or some such name. This document then becomes one of the key inputs to the analysis phase.

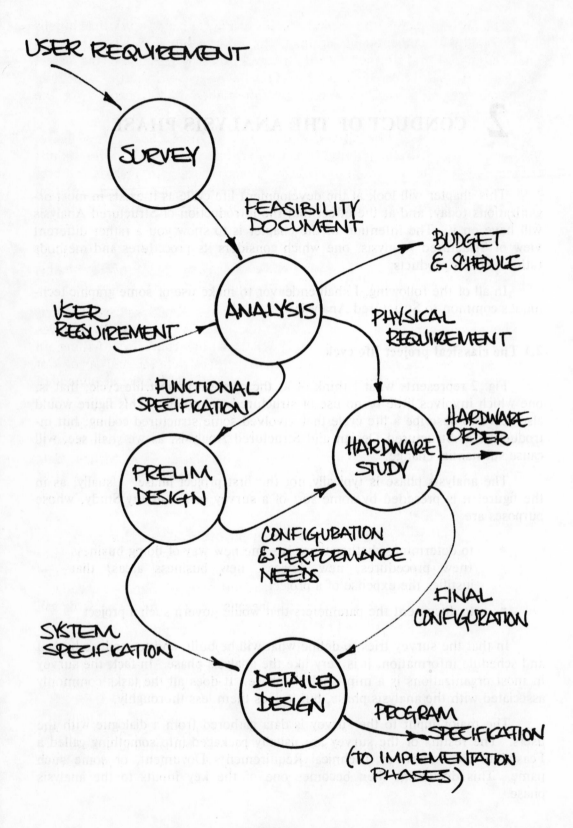

Figure 2

The analysis phase (the second process in Fig. 2) has two principal inputs: the Feasibility Document and, again, direct information flow from the user. These inputs are transformed into the major outputs of the phase, the Target Document (usually referred to as a Functional Specification), budget, schedule, and some sort of physical requirements information that is needed for the hardware study.

I have portrayed the hardware study as a separate phase in my figure, although most organizations lump it in with analysis (incorrectly, to my mind) or with design. The hardware study accepts input from both analysis and preliminary design, and uses it to select new hardware, arrange for augmentation of current hardware, or evaluate the effect of the new system on the current configuration.

The major analysis phase product, the Functional Specification, is passed to preliminary design, the first of the two design phases. Other names for this phase include system design, external design, and philosophical design.

The preliminary design phase uses its input, the Functional Specification, to come up with something called a System Specification. This usually involves selecting modules, drawing a system flowchart or the equivalent, writing up some sort of narrative on each module, and perhaps defining and laying out some of the major data areas and tables. The test plan is frequently generated at this phase.

The System Specification is then passed on to the later design phase, detailed design (or program design). In this phase the nitty-gritty design work is completed. This usually involves internal module design, flowcharting, data definitions, intermodule interface documentation, and the like. The results of the detailed design go to make up the program specification, which is then passed on to the implementation phases.

I have not followed the classical project life cycle beyond this point, but, just for the record, the subsequent phases usually are coding, unit testing, subsystem testing, integration, system testing, acceptance testing, and parallel operation. Overlap among these phases frequently occurs, but it is usually in spite of the company standard rather than because of it — most standards call for a purely linear progression of the implementation phases.

Incidentally, if you looked carefully at Fig. 2 and read all the preceding narrative about it, you have been exposed to a picture and its equivalent thousand words. As you can see, much of my elaboration could have been inferred from the figure.

I have shown the classical analysis phase as though it had only four outputs — the Target Document, physical requirements, budget, and schedule — but it frequently has others. In particular, any of the following might be considered legitimate analysis phase by-products:

- *A very tentative equipment configuration.* It is often necessary to go to this extent to be able to produce budgetary figures.

- *A performance document.* This may be required to justify our expectations of being able to handle the total required throughput within the expense limits set out by the budget.

- *A tentative design.* Some thought about design is necessary during the analysis phase, as part of the continuing feasibility testing that must go on during the specification process. It is reasonable for this to be documented; it is not reasonable for it to be considered a binding statement of design approach.

- *A Request For Proposal (RFP).* If software is to be subcontract-ed, the RFP is usually written during the analysis phase.

- *File and data layouts.* These are sometimes necessary to justify equipment expense in the budget.

- *Resource analysis (e.g., disk and core maps).* Again, this may be necessary for budgetary estimating.

- *Project planning information.* This might include Gantt Charts, Pert Charts, lists of milestones, and interim products.

The modern life cycle

A more modern project life cycle is shown in Fig. 3. Its early phases are like those of the classical version (Fig. 2). But starting with design, structured technologies begin to have an important effect. In place of what used to be called preliminary design, it is now fairly common to find some variant of Structured Design (it might instead be called Composite Design or Top-Down Design). And the later phases make significant use of structured coding, top-down implementation, stub testing, and the like.

Our interest is mainly in the analysis phase, but the move toward this new life cycle is relevant to us because it implies a rather different use of the Target Document. Since Structured Analysis will be producing a new kind of Target Document to replace the classical Functional Specification, it is worth considering how that product of analysis is to be used. What is going to happen in the process called Structured Design (Process 3 of Fig. 3)? How are its in-puts, in particular the Functional Specification, to be transformed into the out-puts, packaged design and test plan?

To answer these questions, we need to look *inside* Process 3. Fig. 4 does just that. It shows the interior of Process 3, a more detailed description of the transformation called "Structured Design." As you can see, it is broken down there into four component parts:

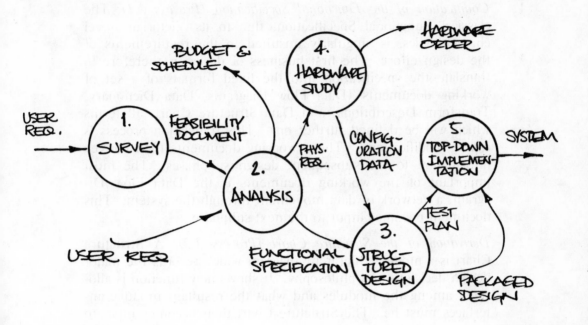

Figure 3

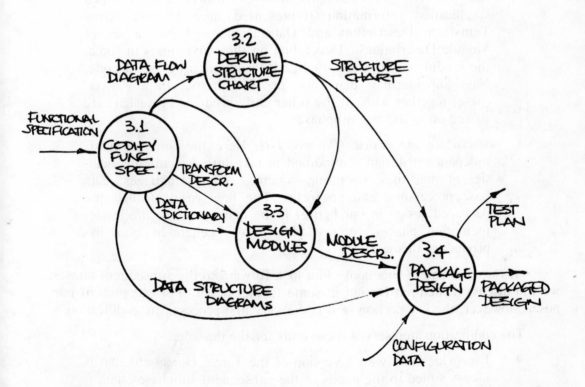

Figure 4

1. *Codification of the Functional Specification (Process 3.1):* The
    classical Functional Specification, due to its Victorian novel
    characteristics, is singularly unsuited to the requirements of
    the design effort. The first business of design is therefore to
    translate the specification into the fixed formats of a set of
    working documents (Data Flow Diagrams, Data Dictionary,
    Transform Descriptions, and Data Structure Chart, all terms
    which will be defined further on). This translation process is
    called "codification." These working documents then convey
    information to the subsequent design subphases. The most
    important of the working documents is the Data Flow Di-
    agram, a network of data movement through the system. This
    document becomes input to the next subphase.

2. *Derivation of the Structure Chart (Process 3.2):* A Structure
    Chart is a modular hierarchy diagram which records the major
    design decisions and philosophy. It shows how function is allo-
    cated among the modules and what the resultant modular in-
    terfaces must be. The Structure Chart then becomes input to
    the next subphase.

3. *Design of the modules (Process 3.3):* With the overall structure
    complete and documented, the internal design of the modules
    can proceed. This usually involves incorporating detailed
    specification information (represented here by the terms
    Transform Description and Data Dictionary) into a set of
    Module Descriptions. Depending on the conventions in force,
    the module descriptions might be documented by pseudocode,
    Nassi-Shneiderman diagrams, or IPO charts. In any case,
    these, together with all the other early products of design, are
    passed on to the last subphase.

4. *Packaging the design (Process 3.4):* Here the environment-
    independent design is modified to take into account the reali-
    ties of machines, operating systems, coding languages, data
    base processors, and so forth. The final result (called the
    packaged design in our figure) is the major input to the imple-
    mentation phases. A test plan is also generated as a by-
    product of this work.

At the very end of this book, in a look forward to the subsequent phases,
we will discuss Structured Design in some further detail. For our present purposes, however, the key portion of it is the very first process, the codification.

The codification step serves these ends for the designer:

-   It provides him with a version of the Target Document that is
    ideally suited to the needs of the subsequent subphases; i.e., it
    is graphic, partitioned, interface oriented, and organized for
    easy updating.

- It helps him to understand and evaluate the Target Document.

- It helps him to check the Target Document for completeness and coherency.

Taken all together, the various components of the codification effort constitute a one-to-one replacement of the Target Document.

2.3 The effect of Structured Analysis on the life cycle

Now let's consider the effect of adding Structured Analysis to the life cycle of Fig. 3. In order to counter the major deficiencies of analysis and make the easiest possible link to design, we are going to

1.   move the codification work back into the analysis phase

2.   make use of the graphics of our codification tools to help us communicate better with the user, which will allow us to . . .

3.   remove redundancy from the Target Document (to make it easier to keep up-to-date)

4.   remove narrative text from the Target Document, replacing it with a more formal equivalent

5.   remove physical information from the Target Document, making it totally logical

Finally, we cross our fingers and try to remove the classical Target Document altogether, i.e., let the codified variant serve our entire need. Of course, we won't always get away with this. Some standards organizations insist on a certain tonnage of text before they'll consider the analysis phase complete.

The modern project life cycle, with Structured Analysis in place, is shown in Fig. 5. Viewed at this level, the only discernible difference from the classical analysis phase is the nature of the Target Document produced. In the case of Structured Analysis, it is a Structured Specification.

There are some further differences that show up when you look into the details of how the new analysis phase is conducted. In order to see these, we have to look *inside* Process 2 of Fig. 5. There we will learn more about how the inputs to analysis (User Interface and Feasibility Document) are transformed into its products (the Structured Specification and the Physical Requirements Data). Fig. 6 shows the interior of Process 2. This figure is a procedural definition of Structured Analysis.

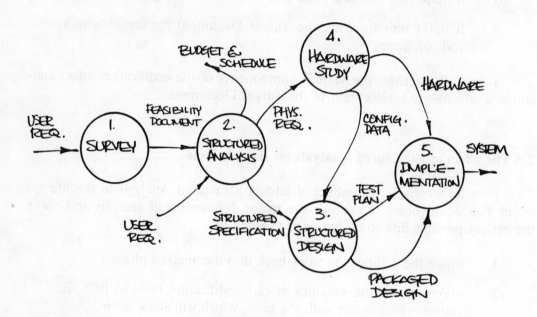

Figure 5

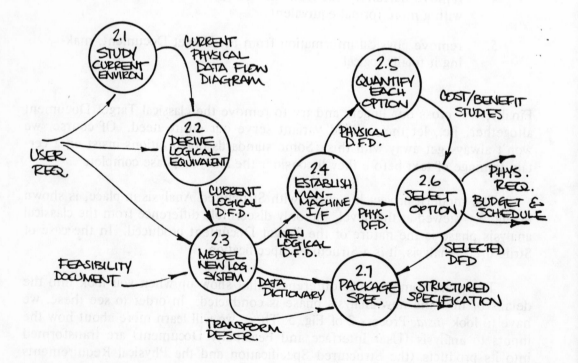

Figure 6

2.4 Procedures of Structured Analysis

As defined by Fig. 6, Structured Analysis is seen to consist of seven component studies:

- Study of the current physical environment, resulting in its documentation by a Current Physical Data Flow Diagram.

- Derivation of the logical equivalent of the current environment, resulting in a Current Logical Data Flow Diagram.

- Derivation of the new logical environment, as portrayed by the New Logical Data Flow Diagram plus supporting documentation.

- Determination of certain physical characteristics of the new environment to produce a set of tentative New Physical Data Flow Diagrams.

- Quantification of cost and schedule data associated with each of the possibilities represented by the set of New Physical Data Flow Diagrams.

- Selection of one option, resulting in one selected New Physical Data Flow Diagram.

- Packaging of the New Physical Data Flow Diagram and supporting documents into the Structured Specification.

Each of these subphases makes use of the concept of the Data Flow Diagram. Although a formal description of the Data Flow Diagram and its attributes is still to come, you are already familiar with its use: All the figures in this chapter are Data Flow Diagrams. As we shall see, the figures used to describe the project life cycle are "leveled Data Flow Diagrams." More about that in just a bit. First I would like you to consider what is involved in each of the subphases of Structured Analysis.

2.4.1 Study of the Current Environment

The first subphase of Structured Analysis, Process 2.1 of Fig. 6, is a complete study of affected user areas. Before this process can begin, you must make an early guess at the scope of the project. This is necessary to determine who the users are, and how much of their work may be subject to change if the new system is introduced. Aside from this, the study of the current environment ought to proceed without much awareness of the impending change (i.e., the change introduced in the Feasibility Document).

Working closely with the users, you learn and document the way things currently work. Rather than do this from the point of view of any one user or set of users, you attempt to assess operations *from the viewpoint of the data.*

This approach will make it easier to draw up the Data Flow Diagram, since a Data Flow Diagram is actually a portrayal of a situation from the data's point of view.

Remember that you are attempting to build a verifiable model of the current environment. In order to make it verifiable, something that the user can understand and validate, you have to make it reflect *his* concept of current operations. So you use his terms and, in some cases, his partitioning. That means that your Data Flow Diagram will be full of department and individual names, form numbers, and the like. All of these serve as checkpoints for the user. They help him to correlate between the paper model you are building and the business area it represents. Since most of these user checkpoint items are physical in nature, the resultant product is termed the "Current Physical Data Flow Diagram."

This study is considered done when we have a complete Current Physical Data Flow Diagram describing the area, and the user has accepted it as an accurate representation of his current mode of operation. The following is a checklist of tasks that make up the current physical study:

- determination of the context to be studied
- identification of users affected (remember to look for all three levels of user)
- user interviews
- Data Flow Diagramming
- collection of sample data types, files, and forms to complement the Data Flow Diagrams
- walkthroughs with the user to verify correctness
- publication of the resultant documentation

2.4.2 Deriving Logical Equivalents

The next step (Process 2.2) is to "logicalize" our model of the current environment. This is largely a cleanup task, during which we remove the physical checkpoint items, replacing each one with its logical equivalent. A particular implementation of policy is replaced by a representation of the policy itself. The underlying objectives of the current operation are divorced from the methods of carrying out those objectives. As an example of the "logicalization" process, consider the two Data Flow Diagram segments shown in Figs. 7 and 8. Fig. 7 is a physical Data Flow Diagram. It shows how a specific report (Report B21) flows through a current operation. Fig. 8 is the logical equivalent.

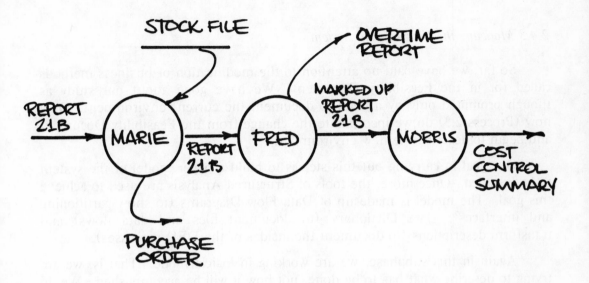

Figure 7

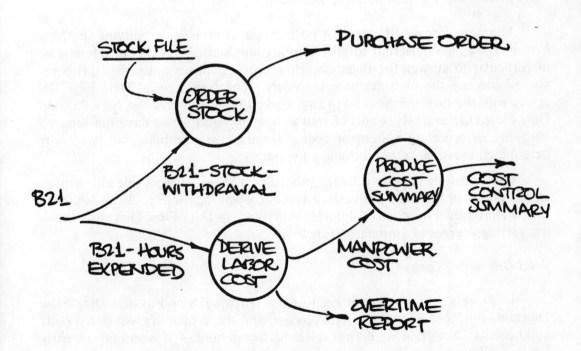

Figure 8

This subphase is complete when the Current Logical Data Flow Diagram is drawn, walked through and verified by the user, and published.

2.4.3 Modeling the New Logical System

So far, we have paid no attention to the modification of business methods called for in the Feasibility Document. We have gone about our study as though prompted only by a need to document the current environment. Only now (Process 2.3) do we incorporate the change from the Feasibility Document and begin to describe the new environment.

Our goal in carrying out this step is to build a paper model of the system to be installed. Once more, the tools of Structured Analysis are used to achieve this goal. The model is made up of Data Flow Diagrams (to show partitioning and interfaces), Data Dictionary (to document files and data flows) and transform descriptions (to document the insides of the DFD processes).

Again in this subphase, we are working in logical mode. That is, we are trying to describe what has to be done, not how it will be accomplished. We do not even distinguish between those procedures that will be automated and those which will be manual. We simply declare the work that must be done, the rules that govern its doing and interfaces among component parts of the whole.

2.4.4 Establishing the Man-Machine Boundary

The next process (Process 2.4 on our figure) involves modifying the New Logical Data Flow Diagram to take into account some physical considerations, in particular to answer the question, How much shall we automate? This involves selecting the man-machine boundary. The New Logical Data Flow Diagram with the man-machine boundary marked on it becomes the New Physical Data Flow Diagram. (It is not, of course, very physical — we have not selected hardware or operating system or coding language or terminals; we have only determined the *scope* of our automated system.)

Good analysis practice calls for producing a menu of workable alternatives at this point, rather than a take-it-or-leave-it single possibility. In order to do this, we produce a number of differently marked up Data Flow Diagrams showing varying degrees of automated function.

2.4.5 Quantifying Options

In Process 2.5, we look at each of the tentative New Physical Data Flow Diagrams produced by the previous process, and try to quantify associated costs and benefits. Note that we do not select hardware here — it is still far too early for that. We select only a genre of hardware for each option, i.e., minicomputer vs. maxi, or on-line facilities vs. batch. Rather than pick a particular vendor's cost figures, we try to stay high so that future hardware selection will not be preempted. It is important that hardware selection be allowed to wait until the beginning stages of design have gotten underway, because only then will we have all the necessary selection criteria established.

How you quantify your options will depend somewhat on your organization and the scope of your project. At the very least, you will have to come up with cost and schedule numbers for each possibility. In addition, you might have to consider

- risk

- financial terms (lease vs. purchase vs. rental)

- air conditioning requirements

- floor space requirements

- maintenance requirements

- operating costs

- training costs

- data conversion costs

- personnel considerations

- prestige, public image, and pizzazz

2.4.6 Packaging the Specification

I bypass the actual selection process (Process 2.6) in this discussion because, as analysts, we are usually left out of it. Even when we do get to participate, it seems to be more of a political process than a technical one (dealing, for instance, with the "Only IBM can save us" and "IBM over my dead body" factions that exist in any organization).

When the selection process is over, all that is left is a minor packaging task (Process 2.7). We may want to do some redrafting and general prettying up of the components of the Structured Specification.

2.5 Characteristics of the Structured Specification

The final result, the Structured Specification, consists of an integrated set of

- *Data Flow Diagrams,* showing the major decomposition of function, and all the interfaces among the pieces

- *Data Dictionary*, documenting each of the the interface flows and data stores on any of the Data Flow Diagrams

- *Transform Descriptions*, documenting the internals of the DFD processes in a rigorous fashion (usually through the use of Structured English, Decision Tables, and Decision Trees)

If we have gone about it properly, the Structured Specification ought to have all of the following qualities:

1. It should be *graphic*. The Data Flow Diagrams should present a meaningful picture of what is being specified, a conceptually easy-to-understand presentation of the subject matter.

2. It should be *partitioned*. The processes on the Data Flow Diagrams are the basic elements into which the system is decomposed. As we shall see, this partitioning can be done in a top-down fashion so that there is a smooth progression from the most abstract to the most detailed.

3. It should be *rigorous*. The Data Dictionary will provide a rigorous documentation of the interfaces; and the Transform Descriptions, a rigorous specification of process.

4. It should be *maintainable*. Redundancy is minimized and used in a controlled manner. The process of changing the Structured Specification can be tightly controlled.

5. It should be *iterative*. Portions of the Structured Specification have the characteristic that they can be dealt with separately. We can move them back and forth across the user's desk with a short author-reader cycle until they are right. The working documents that the user deals with are actual components of the Structured Specification. When he approves them, they will appear unchanged in the resultant Structured Specification.

6. It should be *logical, not physical*. By eliminating elements that depend upon hardware, vendor, and operating procedure from the Structured Specification, we protect ourselves against changes to the specification caused by changes in physical thinking.

7. It should be *precise, concise, and highly readable*.

2.6 Political effects of Structured Analysis

Structured Analysis is new. It may seem foreign, or at least unfamiliar to management and user personnel. For these reasons, the problems of getting Structured Analysis accepted in your company are not too different from the ones you encountered in trying to get started with Structured Programming or Walkthroughs or Librarians or any of the new techniques.

For a good discussion of this kind of problem, I refer you to Ed Yourdon's *How to Manage Structured Programming* cited in the Bibliography. The points he makes about introducing any new discipline into your methodology are the following:

- Try it first on a small, non-critical pilot project.

- Try it with people who are favorably disposed to the idea; don't cram it down anybody's throat.

- Document the effects in a concise and not overly religious manner. (Getting carried away may create a backlash.)

- Build expertise before trying to implement it on a global basis.

Above and beyond the problem of newness, there are some political considerations which apply specifically to Structured Analysis. These fall into two categories: effect on the life cycle, and effect on the user-analyst relationship.

2.6.1 Effect on the Project Life Cycle

An initial and very disquieting effect of Structured Analysis is that the project life cycle is front-end loaded. The analysis phase takes up a larger percentage of total project manpower. There are three reasons for this:

- Structured Analysis calls for a rigorous study of the user area, a study which was frequently skipped in the classical approach.

- Structured Analysis causes the analyst to do more than specify; he must also partition what he is specifying.

- Structured Analysis results in moving certain tasks that used to be done much later in the project life cycle up into the analysis phase. As we will see, one of these tasks is documentation of user procedures, something that used to be postponed until after system delivery.

Of course, the justification for additional manpower in analysis is that less manpower will be required later on. However, management is inclined to be very suspicious of this. Management is suspicious of all time spent prior to writing hard code. It is management's long-standing view that the project team really uses all the precoding phases to rest up before getting into the real work of building the system.

In order to justify front-loading time, you have to come down hard on the high costs of poor analysis. Such costs may be felt in the form of

- projects that never get finished

- projects that never get started because of lack of agreement on what to do

- projects that deliver unusable products

2.6.2 Effect on the User-Analyst Relationship

When we introduced ideas like recursive testing or pseudocode, the user was usually unaware of any change in our methods. If he did sense a change, it was in an area of activity where he had no great expertise and even less interest. But Structured Analysis is different. When we introduce Structured Analysis into the life cycle, the user is acutely aware that something has changed. The Target Document that he has to look at is different. Some of the skills that have to be built to achieve a successful conversion to Structured Analysis are user skills.

It is certainly possible to sour users on the concept of Structured Analysis by introducing it to them too heavy-handedly. But organizations in all different business areas have managed to convince their users of the value of Structured Analysis methods, and have successfully taught them its basic skills. Keep these ideas in mind when introducing new documents and methods to your user:

- Users are not dummies. Some of the methods they work with are far more complicated than Structured Analysis. Years ago, it was fashionable to look down on users because they had so little knowledge of EDP. Nowadays, the user areas we are automating are much more complex, and the users are correspondingly more high powered.

- Users are inclined to have little patience with jargon. If you refer to one of your drawings as a "Third Normal Form Structured Decomposition Hierarchy," they are liable to decide that they have no business looking at it since they are not experts in such things. If you refer to it as a "Picture of What's Going On Here," they will examine it constructively and help you to get it right.

- Users need have only a reading knowledge of Structured Analysis. There is no need for them to learn how to write Structured Specifications (although many users do become adept at marking them up).

2.6.3 Partitioning of Effort

(healthy)

A last political effect of Structured Analysis is totally salubrious: It provides an effective way of partitioning manpower during the analysis phase. Once the top-level decomposition of function has been worked out, the whole is divided into separable pieces. Manpower can be allocated on the same basis, one analyst or team to each piece. The interfaces among the analysts correspond to the interfaces among the pieces.

This means that, even though proportionately more manpower is used in the analysis phase, the total elapsed time need not be increased.

2.7 Questions and answers

At this point in my seminars I am usually deluged with questions and some objections. Because I can't guess which particular ones are on your mind at the moment, I have reproduced some of the most common ones below, along with my answers.

Doesn't it take forever to do such a complete study of the current environment? Yes. You ought to allow about 30 percent of the total analysis phase manpower for this task.

How can we justify this to our management? Perhaps you can't — some managers can't even be convinced of the necessity for design. They would have you go directly into coding ("If we don't start coding now, we'll never be done on time.") The strongest justification for a complete study of the current environment is that the work has to be done eventually. The choice really boils down to doing it up front where it can be a positive aid to future work, or going ahead with an incomplete understanding of the user area and then fixing things up during acceptance testing. A sizable additional benefit accrues from an early and complete study of the user area: You build legitimacy with the user. When you go to him later on (with your description of the new system) and tell him that you think you have a good idea for how he ought to function in the future, at least he knows you and knows you are competent to discuss new procedures with him. No user is going to be able to accord you any credibility on a new way to do his job if he doesn't think you understand the old way.

Suppose there just isn't time for such a complete study? That's like asking "Suppose there isn't time for design?" I have never seen anyone save time by doing the work in the wrong order, or by skipping required steps. If there isn't time to do it right, there certainly isn't time to do it wrong.

Suppose the new environment is going to be very different from the old? Of course, if the change is so major that the old environment has no relevance at all, there would be no point in studying it. If your company used to be a bank, for instance, and is now going to be a discotheque, this objection would be valid. My experience is that management, in a desperate urge to "get on with the real work," is all too ready to skip the thorough study of the user area on any available pretext. You have to be very hard-nosed about this.

Suppose the physical parameters have been established before the project begins? This does happen a lot. The whole project direction is modified by such an approach. Instead of asking, What changes are called for in the way we do business, and how can we best effect them? the project is left to ask, What can we stuff into that minicomputer that the boss bought on speculation? I believe that the best way to deal with such a situation is to ignore it until you have completed your study of the user area.

You have shown the analysis phase proceeding in a linear fashion; isn't there feedback among the various subphases? Yes, absolutely. There is a particularly important possibility of feedback between Process 2.4 and Process 2.3 in Fig. 6. This involves going back to logical dataflowing to incorporate features that are purely dependent upon the physical option selected.

Are people currently using Structured Analysis? Yes. Structured Analysis has been in use since the early 1970's. In a recent survey of readers of *The Yourdon Report*,[1] nearly a quarter of the 126 respondents were making some use of Structured Analysis. Typical users are large institutional companies (banks, insurance companies, government contractors) where analysis problems are the worst, and hence the willingness to try something new is greater.

Don't you have difficulty getting users to work with this new kind of specification? Some. The very fact that it is different is some cause for concern. However, there have been so many successful projects with Structured Analysis that users are inclined to give it a try. Also, the tools of Structured Analysis are not unfamiliar to users — Data Flow Diagrams, for example, are adapted from long-existing tools of manual systems analysis. Surprisingly, the major objections seem to come from standards organizations, or from the EDP side, rather than from the users. I will have some things to say later about introducing the ideas of Structured Analysis to your users — about how to get them hooked.

---

[1] 1977 Productivity Survey, *The Yourdon Report*, Vol. 2, No. 3 (March 1977).

3 THE TOOLS OF STRUCTURED ANALYSIS

The purpose of this chapter is to give you a look at each one of the tools of Structured Analysis at work. Once you have a good idea of what they are and how they fit together, we can go back and discuss the details.

3.1 A sample situation

The first example I have chosen is a real one, involving the workings of our own company, Yourdon inc. To enhance your understanding of what follows, you ought to be aware of these facts:

1. Yourdon is a medium-sized computer consulting and training company that teaches public and inhouse sessions in major cities in North America and occasionally elsewhere.

2. People register for seminars by mail and by phone. Each registration results in a confirmation letter and invoice being sent back to the registrant.

3. Payments come in by mail. Each payment has to be matched up to its associated invoice to credit accounts receivable.

4. There is a mechanism for people to cancel their registrations if they should have to.

5. Once you have taken one of the company's courses, or even expressed interest in one, your name is added to a data base of people to be pursued relentlessly forever after. This data base contains entries for tens of thousands of people in nearly as many organizations.

6. In addition to the normal sales prompting usage of the data base, it has to support inquiries such as

    -   When is the next Structured Design Programming Workshop in the state of California?

    -   Who else from my organization has attended the Structured Analysis seminar? How did they rate it?

- Which instructor is giving the Houston Structured Design and Programming Workshop next month?

In early 1976, Yourdon began a project to install a set of automated management and operational aids on a PDP-11/45, running under the UNIX operating system. Development of the system — which is now operational — first called for a study of sales and accounting functions. The study made use of the tools and techniques of Structured Analysis. The following subsections present some partial and interim products of our analysis.

3.2 A Data Flow Diagram example

An early model of the operations of the company is presented in Fig. 9. It is in the form of a Logical Data Flow Diagram. Refer to that figure now, and we'll walk through one of its paths. The rest should be clear by inference.

Input to the portrayed area comes in the form of Transactions ("Trans" in the figure). These are of five types: Cancellations, Enrollments, Payments, Inquiries, plus those that do not qualify as any of these, and are thus considered Rejects. Although there are no people or locations or departments shown on this figure (it is logical, not physical), I will fill some of these in for you, just as I would for a user to help him relate back to the physical situation that he knows. The receptionist (a physical consideration) handles all incoming transactions, whether they come by phone or by mail. He performs the initial edit, shown as Process 1 in the figure. People who want to take a course in Unmitigated Freelance Speleology, for example, are told to look elsewhere. Incomplete or improperly specified enrollment requests and inquiries, etc., are sent back to the originator with a note. Only clean transactions that fall into the four acceptable categories are passed on.

Enrollments go next to the registrar. His function (Process 2) is to use the information on the enrollment form to update three files: the People File, the Seminar File, and the Payments File. He then fills out an enrollment chit and passes it on to the accounting department. In our figure, the enrollment chit is called "E-Data," and the accounting process that receives it is Process 6.

Information on the chit is now transformed into an invoice. This process is partially automated, by the way — a ledger machine is used — but that information is not shown on a logical Data Flow Diagram.

The invoice passes on to the confirmation process (which happens to be done by the receptionist in this case). This task (Process 7) involves combining the invoice with a customized form letter, to be sent out together as a confirmation. The confirmation goes back to the customer.

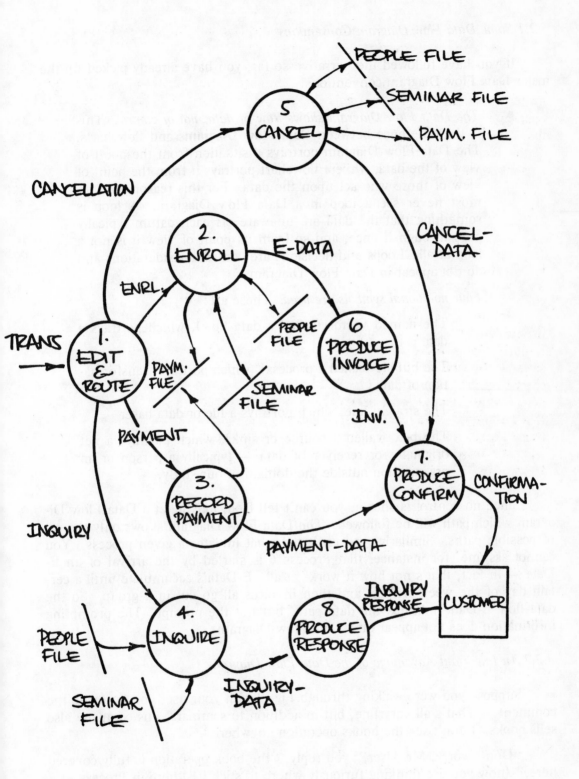

Figure 9

3.2.1 Some Data Flow Diagram Conventions

If you have followed the narrative so far, you have already picked up the major Data Flow Diagram conventions:

- *The Data Flow Diagram shows flow of data, not of control.* This is the difference between Data Flow Diagrams and flowcharts. The Data Flow Diagram portrays a situation from the point of view of the data, while a flowchart portrays it from the point of view of those who act upon the data. For this reason, you almost never see a loop in a Data Flow Diagram. A loop is something that the data are unaware of; each datum typically goes through it once, and so from its point of view it is not a loop at all. Loops and decisions are control considerations and do not appear in Data Flow Diagrams.

- *Four notational symbols are used.* These are:

  - The named vector (called a data flow), which portrays a data path.

  - The bubble (called a process), which portrays transformation of data.

  - The straight line, which portrays a file or data base.

  - The box (called a source or sink), which portrays a net originator or receiver of data — typically a person or an organization outside the domain of our study.

Since no control is shown, you can't tell from looking at a Data Flow Diagram which path will be followed. The Data Flow Diagram shows only the set of possible paths. Similarly, you can't tell what initiates a given process. You cannot assume, for instance, that Process 6 is started by the arrival of an E-Data — in fact, that's not how it works at all. E-Data's accumulate until a certain day of the week arrives, and then invoices all go out in a group. So the data flow E-Data indicates the data path, but not the prompt. The prompting information does not appear on a Data Flow Diagram.

3.2.2 An Important Advantage of the Data Flow Diagram

Suppose you were walking through Fig. 9 with your user and he made the comment: "That's all very fine, but in addition to seminars, this company also sells books. I don't see the books operation anywhere."

"Don't worry, Mr. User," you reply, "the book operation is fully covered here," (now you are thinking furiously where to stick it) "here in Process . . . um . . . Process Number 3. Yes, definitely 3. It's part of recording payments, only you have to look into the details to see that."

Analysts are always good at thinking on their feet, but in this case, the effort is futile. The book operation has quite simply been *left out* of Fig. 9 — it's wrong. No amount of thinking on your feet can cover up this failing. No books flow in or out, no inventory information is available, no reorder data flows are shown. Process 3 simply doesn't have access to the information it needs to carry out books functions. Neither do any of the others.

Your only option at this point is to admit the figure is wrong and fix it. While this might be galling when it happens, in the long run you are way ahead — making a similar change later on to the hard code would cost you considerably more grief.

I have seen this happen so many times: an analyst caught flat-footed with an incorrect Data Flow Diagram, trying to weasel his way out, but eventually being forced to admit that it is wrong and having to fix it. I conclude that it is a natural characteristic of the tool:

> When a Data Flow Diagram is wrong, it is glaringly, demonstrably, indefensibly wrong.

This seems to me to be an enormous advantage of using Data Flow Diagrams.

3.2.3 What Have We Accomplished With a Data Flow Diagram?

The Data Flow Diagram is documentation of a situation from the point of view of the data. This turns out to be a more useful viewpoint than that of any of the people or systems that process the data, because the data itself sees the big picture. So the first thing we have accomplished with the Data Flow Diagram is to come up with a meaningful portrayal of a system or a part of a system.

The Data Flow Diagram can also be used as a model of a real situation. You can try things out on it conveniently and get a good idea of how the real system will react when it is finally built.

Both the conceptual documentation and the modeling are valuable results of our Data Flow Diagramming effort. But something else, perhaps more important, has come about as a virtually free by-product of the effort: The Data Flow Diagram gives us a highly useful *partitioning* of a system. Fig. 9 shows an unhandily large operation conveniently broken down into eight pieces. It also shows all the interfaces among those eight pieces. (If any interface is left out, the diagram is simply wrong and has to be fixed.)

Notice that the use of a Data Flow Diagram causes us to go about our partitioning in a rather oblique way. If what we wanted to do was break things down, why didn't we just do that? Why didn't we concentrate on functions and subfunctions and just accomplish a brute-force partitioning? The reason for this is that a brute-force partitioning is too difficult. It is too difficult to say with any assurance that some task or group of tasks constitutes a "function." In fact, I'll bet you can't even define the word function except in a purely mathematical sense. Your dictionary won't do much better — it will give a long-winded definition that boils down to saying a function is a bunch of stuff to be done. The concept of function is just too imprecise for our purposes.

The oblique approach of partitioning by Data Flow Diagram gives us a "functional" partitioning, where this very special-purpose definition of the word functional applies:

> A partitioning may be considered *functional* when the interfaces among the pieces are minimized.

This kind of partitioning is ideal for our purposes.

3.3 A Data Dictionary example

Refer back to Fig. 9 for a moment. What is the interface between Process 3 and Process 7? As long as all that specifies the interface is the weak name "Payment-Data," we don't have a specification at all. "Payment-Data" could mean anything. We must state precisely what me mean by the data flow bearing that name in order for our Structured Specification to be anything more than a hazy sketch of the system. It is in the Data Dictionary that we state precisely what each of our data flows is made up of.

An entry from the sample project Data Dictionary might look like this:

Payment-Data   =   **Customer-Name +**
                         Customer-Address +
                         Invoice-Number +
                         Amount-of-Payment

In other words, the data flow called "Payment-Data" consists precisely of the items Customer-Name, Customer-Address, Invoice-Number, and Amount-of-Payment, concatenated together. They must appear in that order, and they must all be present. No other kind of data flow could qualify as a Payment-Data, even though the name might be applicable.

You may have to make several queries to the Data Dictionary in order to understand a term completely enough for your needs. (This also happens with conventional dictionaries — you might look up the term perspicacious, and find that it means sagacious; then you have to look up sagacious.) In the case of the example above, you may have to look further in the Data Dictionary to see exactly what an Invoice-Number is:

Invoice-Number   =   **State-Code +**
                          Customer-Account-Number +
                          Salesman-ID +
                          Sequential-Invoice-Count

Just as the Data Flow Diagram effects a partitioning of the area of our study, the Data Dictionary effects a top-down partitioning of our data. At the highest levels, data flows are defined as being made up of subordinate elements. Then the subordinate elements (also data flows) are themselves defined in terms of still more detailed subordinates.

Before our Structured Specification is complete, there will have to be a Data Dictionary entry for every single data flow on our Data Flow Diagram, and for all the subordinates used to define them. In the same fashion, we can use Data Dictionary entries to define our files.

3.4 A Structured English example

Partitioning is a great aid to specification, but you can't specify by partitioning alone. At some point you have to stop breaking things down into finer and finer pieces, and actually document the makeup of the pieces. In the terms of our Structured Specification, we have to state what it takes to do each of the data transformations indicated by a bubble on our Data Flow Diagram.

There are many ways we could go about this. Narrative text is certainly the most familiar of these. To the extent that we have partitioned sufficiently before beginning to specify, we may be spared the major difficulties of narrative description. However, we can do even better.

A tool that is becoming more and more common for process description is Structured English. Presented below is a Structured English example of a user's invoice handling policy from the sample analysis. It appears without clarification; if clarification is needed, it has failed in its intended purpose.

==

POLICY FOR INVOICE PROCESSING

If the amount of the invoice exceeds $500,
    If the account has any invoice more than 60 days overdue,
        hold the confirmation pending resolution of the debt.
    Else (account is in good standing),
        issue confirmation and invoice.
Else (invoice $500 or less),
    If the account has any invoice more than 60 days overdue,
        issue confirmation, invoice and write message on the
        credit action report.
    Else (account is in good standing),
        issue confirmation and invoice.

==

3.5 A Decision Table example

The same policy might be described as well by a Decision Table:

		RULES		
CONDITIONS	**1**	**2**	**3**	**4**
1. Invoice > $500	Y	N	Y	N
2. Account over- due by 60 + days	Y	Y	N	N
ACTIONS				
1. Issue Confirmation	N	Y	Y	Y
2. Issue Invoice	N	Y	Y	Y
3. Msg to C.A.R.	N	Y	N	N

3.6 A Decision Tree example

As a third alternative, you might describe the same policy with a Decision Tree. I have included the equivalent Decision Tree as Fig. 10.

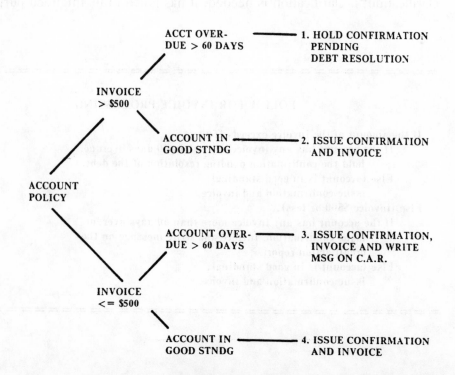

ACTION

Figure 10

PART 2

FUNCTIONAL DECOMPOSITION

4 DATA FLOW DIAGRAMS

One of the key goals of Structured Analysis is to come up with a useful partitioning of the area to be specified. Then, rather than a monolithic Victorian novel specification, we can write an integrated set of mini-specs. We hope to be able to partition down to the point where our mini-specs will be about one page each, so we will be doing a lot of partitioning. Our major partitioning tool is going to be the Data Flow Diagram.

4.1 What is a Data Flow Diagram?

Let's start out with a definition:

A Data Flow Diagram is a network representation of a system. The system may be automated, manual, or mixed. The Data Flow Diagram portrays the system in terms of its component pieces, with all interfaces among the components indicated.

A Data Flow Diagram example, the picture that is worth a thousand words, is shown in Fig. 11.

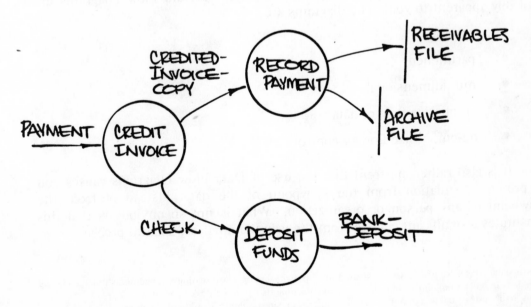

Figure 11

4.2 A Data Flow Diagram by any other name . . .

There are some other terms commonly used by analysts to refer to Data Flow Diagrams, including DFD's, Data Flow Graphs, and bubble charts. In addition, there are some long-established manual systems analysis tools that are not too different, such as paper flowcharts, Document Flow Diagrams, and Petri Networks. Many organizations are surprised to find that their users adapt more readily than their EDP staff to the use of Data Flow Diagrams. In fact, many users have been conversant with such techniques since the 1940's. Their goals in using these classical network approaches were different from the set of Structured Analysis objectives we established earlier.[1] But the techniques themselves look quite familiar to the user community.

A slight variation on the idea of a Data Flow Diagram is something called SADT. This is a proprietary version of the concept, developed by Softech, Inc., a consulting company that did some of the earliest work in Structured Analysis. A set of papers documenting their technique, published in a special Structured Analysis issue of the *IEEE Transactions on Software Engineering* (cited in the Bibliography), contains some good examples of SADT.[2] Their technique appears somewhat different from the tools described in this book, mostly because it uses square bubbles. I think you will find the similarities much more significant than the differences.

4.3 DFD characteristics — inversion of viewpoint

By now, the most significant characteristics of Data Flow Diagrams are probably apparent to you. The diagrams are

- graphic

- partitioned

- multidimensional

- emphasize flow of data

- de-emphasize flow of control

It is also rather apparent that the use of Data Flow Diagrams causes you to present a situation from the viewpoint of the data, instead of from the viewpoint of any person or organization. What is not so obvious is that this constitutes a significant change from the way that analysis used to proceed.

---

[1] For instance, Petri Networks were first used to help managers organize seating arrangements in large clerical operations in order to minimize time spent moving documents from one desk to another.
[2] This issue also contains a number of other interesting ideas about Structured Analysis. For instance, it presents some of the work done at the University of Michigan by Dr. Daniel Teichroew and his staff on the ISDOS project, an attempt to automate portions of the analysis process.

In classical analysis, we first try to see operations from the user's viewpoint; i.e., we interview him and try to learn from him how things work. Then we spend the rest of our time trying to document the working of modified operations *from the system's viewpoint.* (Notice that this approach is pervasive in unstructured technology; a flowchart, for instance, is design documentation from the system's point of view.)

The inversion of viewpoints occasioned by Structured Analysis is that we now present the workings of a system as seen by the data, not as seen by the data processors. The advantage of this approach is that the data sees the big picture, while the various people and machines and organizations that work on the data see only a portion of what happens. As you go about doing a Structured Analysis, you will find yourself more and more frequently attaching yourself to the data and following it through the operation. I think of this as "interviewing the data." It is usually more productive than any other single interview.

5 DATA FLOW DIAGRAM CONVENTIONS

In adopting our set of Data Flow Diagram conventions, we have attempted to use the established manual system notations (Document Flow Diagrams, Petri Networks, and the like) wherever possible. This will make the Data Flow Diagram look familiar to users of older network technologies, although there are some significant differences.

5.1 Data Flow Diagram elements

Data Flow Diagrams are made up of only four basic elements:

1. data flows, represented by named vectors

2. processes, represented by circles or "bubbles"

3. files, represented by straight lines

4. data sources and sinks, represented by boxes

Fig. 12 is a portion of a Data Flow Diagram that includes each of the four elements. Consistent with our philosophy of concentrating primarily on data flow, we read it as follows:

X's arrive from the source S and are transformed into Y's by the process P1 (which requires access to the file F to do its work). Y's are subsequently transformed into Z's by the process P2.

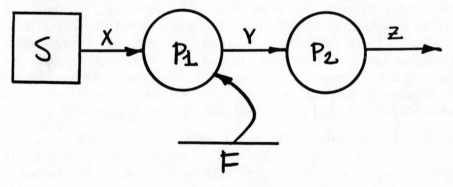

Figure 12

The processes P1 and P2 ought to receive their names from the data flows that move into and out of them; for instance P2 should be named "Transform Y's into Z's." The opposite convention would have us concentrate first on the processes; in this case, we would adapt data flow names from the processes — Y would be called "Stuff generated by P1 for the use of P2." I think you can see that this might encourage partitioning with more complicated interfaces.

Let's look at the four DFD elements one by one.

5.1.1 First DFD Element: The Data Flow

A data flow portrays some interface among components of a Data Flow Diagram. Most data flows move between processes, but they can just as well flow into or out of files, and to and from destination boxes and source boxes, respectively. In all cases, we use the notational symbol of a named vector to show the interface.

Consider the segment shown in Fig. 13. Suppose we know (from our Data Dictionary) that the data flow called "Payment" always consists of the customer's pink copy of the invoice plus a check. Why have we chosen to show Payment here as a single data flow rather than two? What is a data flow? To answer these questions we need to define data flow in a way that distinguishes between the interface and the information that passes over the interface:

> A *data flow* is a pipeline through which packets of information of known composition flow.

The packet of information in our example is composed of a check and a pink invoice copy. The packet flows from the Customer to the process Credit Invoice over the pipeline called Payment.

In Fig. 14 we see a Data Flow Diagram segment that has two separate data flows moving between two processes. The reason for this is that the different items flowing over them do not constitute a packet; they never travel together. (Stock-Report, for instance, might be generated on a weekly basis, while Emergency-Reorder-Ticket might be generated whenever a critical shortage of the item is discovered.) They have different purposes, are probably produced by different portions of the source process, and generally don't have much to do with each other. For these reasons, they require separate pipelines.

If we forced them through the same pipeline, we would only obscure the interface. We couldn't even come up with a very satisfactory name for the combination; we would probably call it something wishy-washy like Stock-Data or Stock-Control-Output. (Of course, you could always refer to the Data Dictionary to see the odds and ends that traveled over the interface, but you couldn't tell much by glancing at the diagram.)

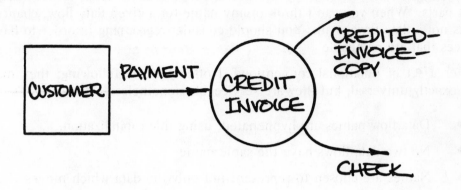

Figure 13

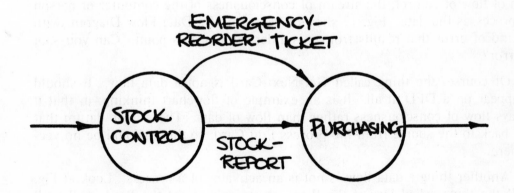

Figure 14

This concept of nameability turns out to be a valuable feedback mechanism to judge how effectively the system is being broken down into its component parts. When you can't think of any name for a given data flow, chances are it is not a data flow at all. You should consider regrouping in order to have interfaces that are nameable.

For a set of notational conventions, I offer you the following; they may not be exactly universal, but they are serviceable nonetheless.

- Data flow names are hyphenated, using title capitalization.

- No two data flows have the same name.

- Names are chosen to represent not only the data which moves over the pipeline, but also what we know about the data. For instance, Fig. 15 shows two data flows, Account-Number and Valid-Account-Number, which are identical in composition. But we know more about the second one, and that additional understanding of it is reflected in its name.

- Fig. 16 shows some further DFD gyrations that you will encounter, data flows that converge and diverge.

- The data flows moving into and out of simple files do not require names — the file name will suffice to describe the pipeline. All other data flows must be named.

A few words about what a data flow is not: A data flow is not a representation of flow of control, the stream of consciousness of the computer or person that processes the data. Fig. 17 shows a segment of a Data Flow Diagram with the kind of error that results from misunderstanding this point. Can you spot the error?

Of course, the thing called Get-Next-Card is not a data flow. It should not appear on a DFD at all. It is an example of flowchart thinking, in that it portrays flow of consciousness rather than flow of data. There is no datum that flows back to the source from the process Edit Card, so there should be no vector there.

Another thing a data flow is not is an activator of a process. Look at Fig. 18. If the thing called Day-of-the-Week exists only to get the Produce Payroll process started, then we should think of it as a control item and not show it as a data flow.

How shall we represent flow of control and movement of control items on the DFD? We shall not. In fact, we will defer specifying this information as long as possible — until generation of the mini-spec. It is procedural in nature, and thus has little to contribute to the big-picture purpose of the Data Flow Diagram.

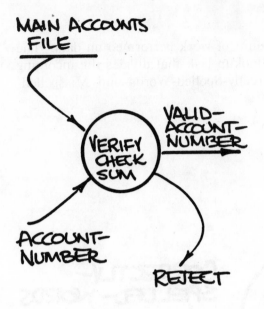

Figure 15

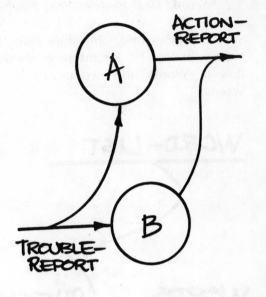

Figure 16

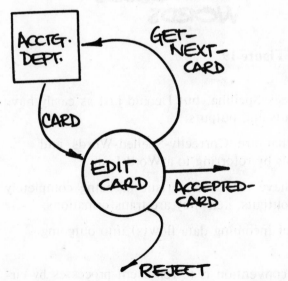

Figure 17

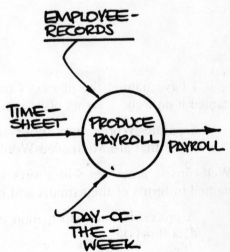

Figure 18

5.1.2 Second DFD Element: The Process

Processes invariably show some amount of work performed on data. The process in Fig. 19, for instance, shows a lookup task that divides the incoming flow of Words into two pipelines, Correctly-Spelled-Words and Misspelled-Words.

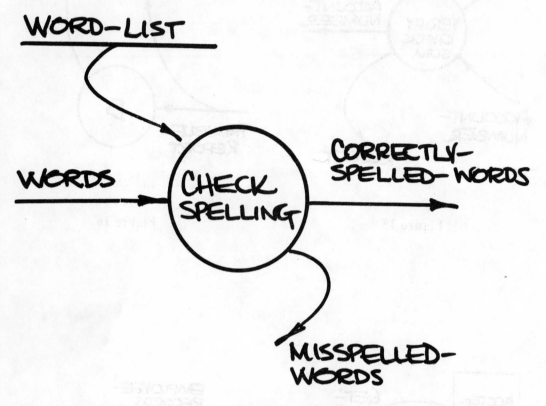

Figure 19

I have named the process Check Spelling, but I could just as easily have named it entirely ⸱ terms of its inputs and outputs:

Separate Words into those that are Correctly-Spelled-Words and those that are Misspelled-Words by referring to a Word-List.

Well-chosen processes will always have the characteristic of being completely named in terms of their inputs and outputs, i.e., they are transformations:

✓ A *process* is a transformation of incoming data flow(s) into outgoing data flow(s).

The most common notational convention is to represent processes by circles (bubbles) on the Data Flow Diagram. Some people use oval bubbles; SADT and certain other Structured Analysis conventions use square bubbles.

Regardless of its shape, each bubble needs a descriptive name. Of course, we are going to specify what the process consists of in much more detail (in the Structured English mini-spec, for instance), but the process name must give the user a general idea or we will not have succeeded in conveying the big picture with the DFD.

In a completed set of Data Flow Diagrams, each process will be given a unique number. The numbering convention will depend on how the various diagrams interrelate, so I will put off talking about it for the moment.

5.1.3 Third DFD Element: The File

Of course you know what a file is. But for the time being, I would like to divorce the concept from connotations that are typically associated with the word. For our purposes:

A *file* is a temporary repository of data.

It may be a tape, or an area of disk, or a card data set, or an index file in someone's drawer, or the little book of deadbeat cardholders that the credit card companies issue from time to time. It might even be a wastebasket. As long as it is a temporary repository of data, it qualifies as a file. The notational convention used to represent a file on the DFD is trivial: a straight line with the file's name in close proximity.

Data bases qualify as files under this definition. The term data base carries along with it even more connotations about physical implementation than a file does. Again, we are going to defer worrying about these physical attributes until much later in the analysis phase.

A comment about file names: Look back at Fig. 19, where the file called Word-List is accessed by the process Check Spelling. It goes without saying that we need to specify the Word-List by more than its name — we will do that in the Data Dictionary — but a well-chosen file name is a great aid to the DFD's readability. Your conceptual understanding of that DFD segment, together with its use of the file, is probably already complete and accurate. (You would certainly be offended if the Word-List file had things in it like overdue payables and the price of eggs.) Some standards organizations, in their attempt to use unique names company wide, have decreed that files ought not to have names like Word-List, but should instead be referred to as BZZX1977P4.IPCRESS2 or some such obscure encoding. Your users will generally not be concerned with, or even aware of, such naming schemes. To make DFD's meaningful to your users you must avoid coded names. Use names that make sense on each DFD. If you have to fight your standards group for this freedom, you have my sympathy. But it is a fight worth fighting.

The direction of arrows leading to or from a file is significant. Fig. 19, for instance, shows data moving out of the file only. The process called Check Spelling does not write onto the file — that's what the DFD says. If Check Spelling is now made into a learning procedure, one that does make new entries in the Word-List under certain circumstances, the figure is no longer correct. It has to be modified to show two-way access. Usually this is done with a double-headed arrow.

Look at Fig. 20. It shows data flowing into the Parts Master File, but no data flowing back to the process. You might wonder how on earth the process can ever do an update without first reading the file. Of course, it cannot. However, Fig. 20 is correct as it is shown. The rule is that *only net flow is shown* to and from the file. If input is required only to be able to do output, then net flow is clearly out. If, on the other hand, some of the information on the file goes into the New-Parts-Report, then the figure would have to be corrected to show net flow in both directions.

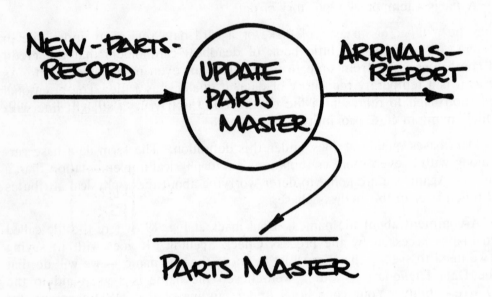

Figure 20

5.1.4 Fourth DFD Element: The Source or Sink

Any system or business area can be described on a DFD with data flows, processes, and files. Sometimes, however, you can substantially increase the readability of your diagram by showing where the net inputs to the system come from and where the net outputs go to. For this we use source/sink boxes, as shown in Fig. 21.

A *source or sink* is a person or organization, lying outside the context of a system, that is a net originator or receiver of system data.

The key qualifier is "lying outside the context of a system." A person or organization inside the context of the system (i.e., within the domain of our study) would be characterized by the processes he or it performs.

The dotted line of Fig. 21 shows the system context. Note that it neatly divides the sources and sinks of system data from the rest of the DFD.

By convention, sources and sinks are represented on a DFD by named boxes. There is no rule against data flowing both into and out of a box. Boxes are used rather sparingly in Data Flow Diagrams and usually not in a very rigorous fashion. Since they represent something outside the area of our major concern, they exist only to provide commentary about the system's connection to the outside world.

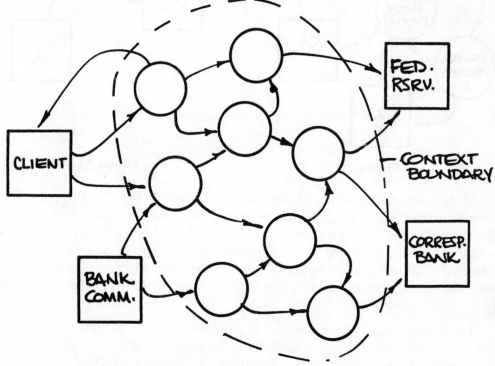

Figure 21

All along I have been using the term "system" without carefully defining it. System, as I use the word, refers to a set of procedures, *both automated and manual,* used to effect a desired end.

With this in mind, look at the system in Fig. 22. The context of this system includes both automated and manual processes. In fact, it includes everything with which we are concerned. What the net sources and sinks do with that data cannot affect the system in unexpected ways — that's why we placed them outside the context of our study.

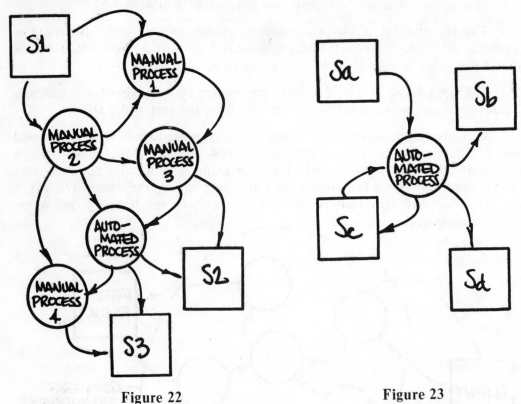

Figure 22   **Figure 23**

If we used source and sink notation as shown in Fig. 23, the preceding comment would not apply. If we are trying to build the computerized procedure called Automated Process in that figure, then the figure implies that we have gone about it all wrong — our analysis was much too restrictive in that it considered a context that consisted of only the Automated Process itself. Obviously, the analysis must include much more than that, or we will be flying blind.

I will come back to the concept of original selection of context. For the present, consider the context of analysis to be whatever we propose to change, plus enough surrounding procedure to buffer the changed area from activity on the outside. Sources and sinks lie outside that context.

5.2 Procedural annotation of DFD's

Fig. 24 shows a Data Flow Diagram with some procedural information added. Specifically, it says that the two incoming data flows must both be present
in order for the process to do its work, and one or the other but not both of the
output data flows will result.

  *   means AND. It portrays *conjunction* of data flows.

  ⊕   means OR. It portrays *disjunction* of data flows.

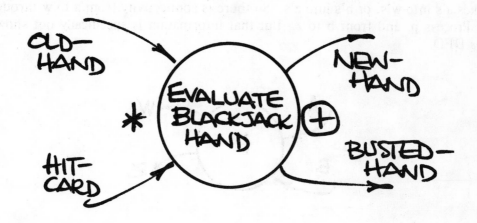

Figure 24

Now that you know the convention, try not to use it at all or use it only
sparingly. It is shown here because you are quite likely to run across it in your
reading on Structured Analysis, but I think the use of procedural annotation
ought to be discouraged. Procedural characteristics are, after all, the very kinds
of things that we can most safely defer. Effort dedicated to them during
analysis is premature. A further mark against procedural annotation is that
people get carried away with it (since it is easy and mechanical); they are forever inventing new ways to annotate more and more complicated interdependen-
cies.

Suppose that Fig. 24 was unadorned with procedural symbols. What
would it mean then? The proper way to read it without the symbols is that
there are two paths in and two paths out of the process — you can't tell anything from the DFD about how they interrelate. Since our Data Flow Diagramming is trying to effect a useful partitioning, and usefulness is dependent on
complexity of interfaces, I suggest that the procedural relationships ("one or
both or sometimes all but never neither") is not terribly relevant to us just
now. The very existence of the path determines how functional the partitioning
will be.

Eventually we are going to write a mini-spec for each and every process on our DFD. As you might guess, the mini-spec for each process will have to specify the relationships that apply among the data flows. The completed Structured Specification, when it is assembled, will consist of the mini-specs plus the DFD's and Data Dictionary. So annotations on the DFD can only increase redundancy — just what we are trying to avoid.

As a final comment about procedural information on a DFD, I point out that a much more critical procedural item is not shown: the relationship between input and output data flows. Imagine a process like the one shown in Fig. 25. We might know that the process does one of two things: It either makes a's into w's, or b's into z's. So there is connectivity from a to w through the Process p, and from b to z. But that information is specifically not shown on a DFD.

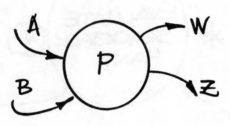

Figure 25

5.3 The Lump Law

By now you must have observed a pattern in the Structured Analysis approach, a pattern of deferring certain kinds of considerations, especially those that are in some sense "physical." I hope you understand that I don't advocate this because of a lack of respect for thoroughness. Any clear and systematic approach to a problem must take great pains to defer thinking about certain details, or everyone involved will quickly be sidetracked by matters of stunning insignificance. Jerry Weinberg's *Introduction to General Systems Thinking,* referenced in the Bibliography, expresses this idea as The Lump Law:

"If we want to learn anything, we mustn't try to learn everything."

at least not right at the beginning.

In choosing what to defer, I propose this very simple rule: *Defer anything you can get away with.* In other words, it is reasonable to defer thinking about those aspects which experience has shown you can safely put off. Definitely included in this category are implementation-related details, what I have been calling physical considerations. These can, for the most part, be ignored during much of the analysis phase. I'm sure you could dream up some exceptions, but let's be sure to think of them as just that, exceptions.

6 GUIDELINES FOR DRAWING DATA FLOW DIAGRAMS

Enough said about DFD characteristics, intentions, and conventions. How do you draw one? As you become adept at it, you will find that different problems call for different methods, but the following is a good general-purpose approach:

1. Identify all net input and output data flows. Draw them in around the outside of your diagram.

2. Work your way from inputs to outputs, backwards from outputs to inputs, or from the middle out.

3. Label all the interface data flows carefully.

4. Label the bubbles in terms of their inputs and outputs.

5. Ignore initialization and termination.

6. Omit details of trivial error paths (for now).

7. Do not show flow of control or control information.

8. Be prepared to start over.

The sections below elaborate on each of these guidelines.

6.1 Identifying net inputs and outputs

The business of determining net inputs and outputs is closely tied to the decision on what the context of the study shall be. Typically, that decision is made at the beginning of the analysis phase, often with very little thought. Selecting the context is a matter of judgment and feel. Regardless of how you go about it, however, the goal is the same: to select a context that is large enough to include everything relevant to the development effort, but small enough to include little or nothing that is irrelevant.

In selecting the context, keep in mind that everything left outside is going to be ignored forever after. By leaving it out of the context, we decide that we will dedicate no analysis resources to studying it. So, if you are in doubt, you should err on the conservative side by including too much in the domain of your study.

Having decided on a boundary, you must now look for data flows that cross it. These are the net inputs and outputs. Write them down on the periphery of your DFD. Note that it is the set of data flows that crosses it that will be our documentation of the context boundary.

Don't be too worried about completeness at this point. Data Flow Diagrams have some self-checking mechanisms that help to track down forgotten data flows.

6.2 Filling in the DFD body

On nearly any project, your first use of Data Flow Diagrams will be to document the current user area. In doing this, you are not called upon to describe how things ought to be but rather how they are. Consequently, you can pattern your DFD on what you see and what the user tells you. Your data flows will be packets of data of which the user is aware, usually things for which he has names. Your processes will be segments of the work going on in his area.

Concentrate first on the data flows. Look for the major pipelines of data moving about the operation. Suppose you identify a significant set of information that the user treats as a unit (i.e., it arrives together, it is processed together, and he thinks of it as a whole). That is certain to end up as a data flow on your DFD. Enter it on your diagram, and try to connect it with the data flows on the periphery. Put bubbles wherever some work is required to transform one data flow or set of data flows into another. Don't worry yet about naming these processes; leave them blank.

Look inside the blank processes that are starting to appear on your DFD. Can you imagine any internal data flows that might be used within the process? Check with the user to see if they do indeed exist. If so, replace the single process by two (or three or four) and put the identified data flow between them.

For each data flow, ask the question: What do I need in order to *build* this item? If it is a Passenger-Manifest, for example, you might determine that it is made up of Collected-Airline-Tkts plus the Original-Flight-List plus however many Standby-Forms there might be. Where do these components come from? Can any of the incoming data flows be transformed into any of these? What intervening processes will be required to effect the transformation?

Enter files on your DFD to represent the various data repositories that the user tells you about. Make sure you know the contents of each file in enough detail that you can figure out flow into and out of it.

Be prepared to go back and modify the context boundary. You may have forgotten an input that is required as a key component of one of your data flows; add it in. You may have an incoming data flow that vanishes — is of no use to anything inside the context; take it out. You might even have a situation like the one in Fig. 26, where your context includes not one but two disconnected networks, one of which is completely separated from the true domain of your interest. In such a case, you can safely eliminate the entire extra network.

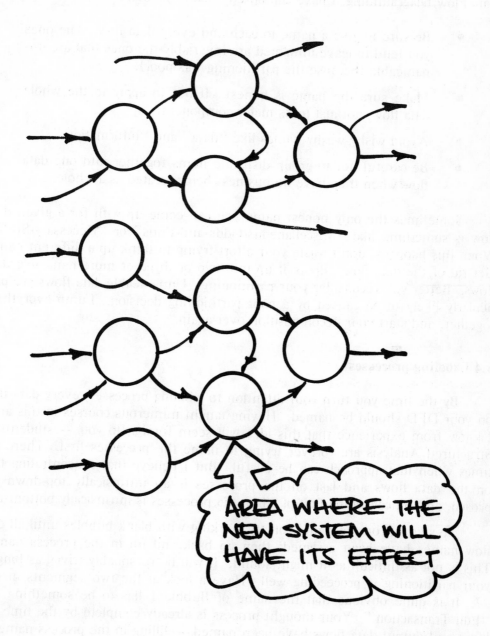

AREA WHERE THE NEW SYSTEM WILL HAVE ITS EFFECT

Figure 26

6.3 Labeling data flows

The names you select for your data flows will have a strong effect on the readability of your DFD. From watching analysts struggle with this aspect of Data Flow Diagramming, I have composed a list of pointers:

- Be sure to give a name to each and every data flow. The ones you tend to leave unnamed are invariably the ones that are un-nameable (because the partitioning was poor).

- Make sure the name is honest. It has to apply to the whole data flow, not just to its major component.

- Avoid wishy-washy names like "data" and "information."

- Be careful not to group disparate items together into one data flow when they have no business being treated as a whole.

Sometimes the only honest name you can come up with for a given data flow is something like "Miscellaneous-Odds-and-Ends" or "Necessary-Stuff." When this happens, don't waste your effort trying to think up a different name. Get rid of the data flow. Break it up into two or three or more nameable data flows. Better yet, reconsider your partitioning. Unnameable data flows are particularly likely to be caused by a poor partitioning decision. Lump everything together, and start your decomposition over again.

6.4 Labeling processes

By the time you turn your attention to labeling processes, every data flow on your DFD should be named. Having taught numerous courses in this area, I know from experience that this idea will seem foreign to you — students of Structured Analysis are forever trying to name the processes first. There are times when that approach can be useful. But I believe that concentrating first on the data flows and last on the processes is an intrinsically top-down approach, and concentrating first of all on the processes is intrinsically bottom-up.

To start off on the right foot, try working with blank bubbles until all data flow names have been assigned; then go back and fill in the process names. This is not as difficult as it might sound. It will be reasonably trivial as long as your partitioning is proceeding well. Have a look at the two segments in Fig. 27. It is quite obvious that the name of Bubble 1 has to be something like "Edit Transaction." Your thought process is already complete by the time the input and output data flows have been named — filling in the process names is mechanical. The opposite is not true at all. Trying to name the data flows a, b, and c associated with the bubble "Stock Control" is an impossible task. None of the thinking has been done.

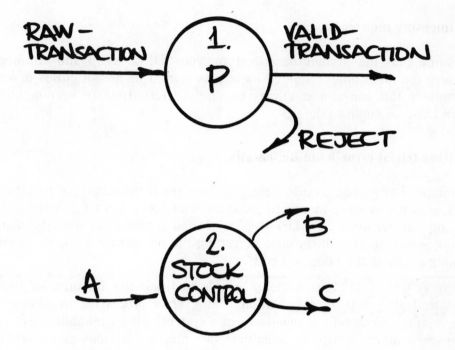

Figure 27

Here is my set of pointers for naming processes:

- Make sure your names are honest. It's unfair to name a process "Edit Transaction" if it also updates the Receivables File, adjusts room temperature, and runs a model railroad.

- Try for names that consist of a single strong action verb and a singular object. If you have two verbs, perhaps you should partition further.

- Beware of wishy-washy words. The worst offenders are words like "process" and "handle" — those words don't mean anything. If the best name you can come up with requires the use of such a word, you have got an unnameable process.

- Repartition to avoid unnameable processes. Break them into two or three parts, or group them with others in order to come up with something you can name easily. If you can't name a process, consider that an indictment of your partitioning effort.

6.5 Documenting the steady state

Assume that the system described by your DFD is up and running. Don't worry for the moment about how it gets started or how it closes down; this is thinking that can almost always be safely deferred, so in keeping with the Lump Law, we ought to defer it.

6.6 Omitting trivial error-handling details

Similarly, I encourage you to defer thinking about the details of trivial error paths. Once you have identified an error data flow, don't feel obliged to follow it any further just yet. Mark it as reject and continue on with the main path. This is consistent with the idea of getting the big picture worked out before looking at any of the odds and ends.

Most systems and most systems people dedicate the majority of their resources to dealing with error situations. Still, error processing usually does not have a strong effect on the *philosophy* of a system. (High-reliability applications like space vehicle control or telephone switching are obvious exceptions to this rule.) Most systems derive their philosophy from the main-line processing. Internal structure must reflect this philosophy. That is the best understood component of the system, and therefore something that should be allowed to shape the thinking of implementation and maintenance personnel.

Can we expect to get away with deferring all consideration of error processing until the later part of analysis? I don't believe so. Here is my rule for deciding which errors shall be allowed to influence our early Data Flow Diagramming: If the error requires no undoing of past processing, ignore it for the moment; if it requires you to back out previous updates or revert a file or files to a previous state, then do *not* ignore it.

I have referred above to "trivial" error paths. I consider an error path to be trivial if it has this characteristic — that no undoing is likely to be required along it.

6.7 Portraying data flow and not control flow

Most people have little trouble breaking old flowcharting habits. If you are ever in doubt about whether one of your data flows might be a control flow in disguise, apply this test: Ask yourself what information flows over the pipeline; if there is none, it is likely to be documenting the stream of consciousness of the data *processor,* rather than the stream of the data itself. Hence it is control flow. Remove it from your Data Flow Diagram.

It is slightly more complicated to eliminate flow of information that is used only for control purposes. The test for this is to ask what the destination process uses the information for. If it is modified and put out as an outgoing data flow or part of one, then it is a legitimate data flow. If it only serves to prompt the process to start doing its work or guide it in how to do its work, then it is control.

Fig. 28 shows some examples of each of the different kinds of false data flows. There are only two legitimate data flows shown in it: Paid-Invoice and Tax-Total.

6.8 Starting over

The human mind is an iterative processor. It never does anything exactly right the first time. It is particularly good at taking an imperfect implementation of a given task, and making significant improvements to it. This it can do over and over again, coming up with better results each time.

Since we work this way quite naturally, why fight it? Be prepared to abandon your early DFD's by replacing them with improved versions. Don't be dismayed, no matter how much you have to throw away. The mechanics of drawing Data Flow Diagrams are trivial — starting from scratch is a small price to pay for a significantly better result. If you find yourself producing a final result on the same sheet of paper you started with, you are not letting your mind work in iterative mode. You ought to be able to point to several generations of improvements.

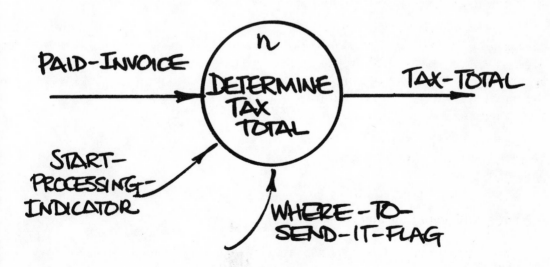

Figure 28

7 LEVELED DATA FLOW DIAGRAMS

Fig. 29 is another example of a Data Flow Diagram, this one describing a simple master file update program. It gives a good overall picture of what goes on in the update, and it also partitions a process that was too big to be handled comfortably as a whole. But this figure points out a potential problem in the use of data flow techniques: It portrays a relatively trivial application, yet it is already about as big a DFD as you can draw on a normal sheet of paper. What are we to do with systems that are orders of magnitude larger? Clearly, we cannot just expand the size of our diagrams to cope with larger and larger requirements — we would quickly find ourselves working on a football field of paper with thousands and thousands of bubbles. We would need a logistics expert to help us find source and destination bubbles for a given data flow. We might even find ourselves committing that Ultimate Crime Against Nature, crossing lines.

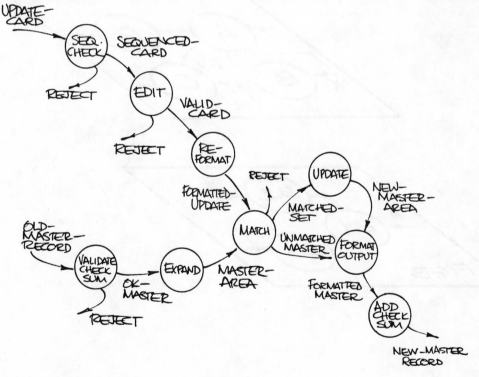

Figure 29

It would be pointless to advocate partitioning tools that were only useful for small applications. After all, it is precisely the largest systems that are most in need of partitioning. In order to use DFD techniques on very large systems, we require the concept of leveling.

7.1 Top-down analysis — the concept of leveling

When a system is too large for its DFD to be shown on a single page, we ought to do an initial partitioning into subsystems. If the subsystems are still too large, we will divide them into sub-subsystems. And so on. Eventually we will end up with components that can be portrayed with simple DFD's of primitive functions.

If we describe each of these successive partitionings with a DFD, we will have a leveled set of Data Flow Diagrams. Fig. 30 shows part of a leveled DFD set. The full set of diagrams will describe our system every bit as rigorously as the football field version would have and will result in the same decomposition. But it will be a good deal more readable.

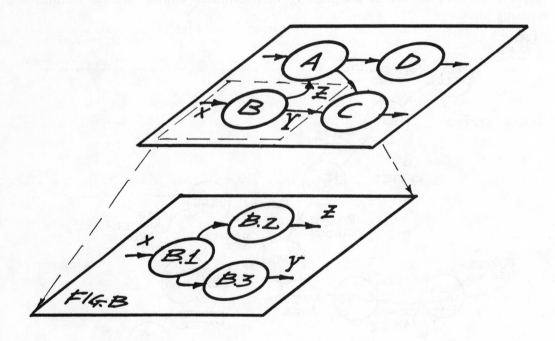

Figure 30

Look at Fig. 31 to see a sample set of leveled DFD's. The whole story is told there. Look at the individual diagrams that make up the set. Try to figure out how they interrelate. What is the convention? I am going to present it as formally as I can in the sections below, but before you get there you should have guessed it all.

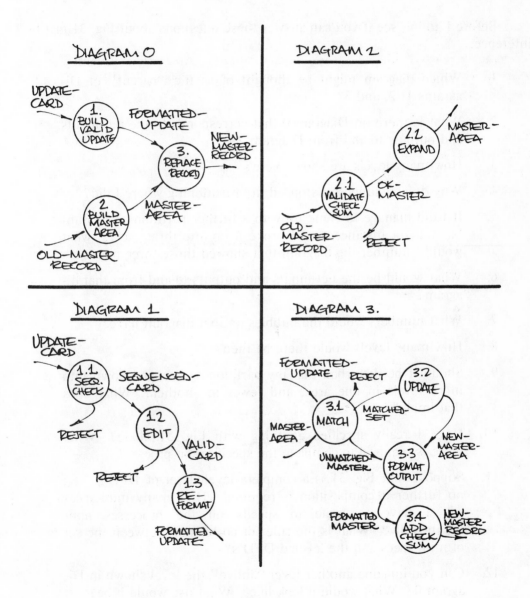

Figure 31

Does the application described in Fig. 31 seem familiar? It is exactly the same as the one presented in Fig. 29, the master file update. Compare the two figures; they are, respectively, a leveled DFD set and its unleveled equivalent.

A Guessing Game

Before I go on, see if you can answer these questions about Fig. 31 just by inference:

1.   Which diagram might be thought of as the "parent" of Diagrams 1, 2, and 3?

2.   What is there on Diagram 0 that corresponds to the net inputs and outputs to and from Diagram 2?

3.   How many levels are shown here?

4.   Why do you suppose I stopped my partitioning where I did?

5.   If I did manage to come up with a further partitioning of Bubble 1.2, for instance, and I broke it up into three pieces, what would I number the diagram that showed those three pieces?

6.   What would be the net inputs and outputs to and from that diagram?

7.   What numbers would the bubbles on that diagram have?

8.   How many levels would there be then?

9.   Should I be disturbed that my partitioning dropped down into more levels in one area, and fewer in another? What does that imply?

10.   Have I really specified anything with Fig. 31 alone? What more is required to complete the specification process?

11.   Suppose that Fig. 31 is a complete leveled set of DFD's, i.e., no further decomposition is required. How many mini-specs will I have to write to specify all the processes non-redundantly? What is the rule for correlation between the set of mini-specs and the leveled DFD's?

12.   Can you imagine another level "above" the level shown in Diagram 0? What would it look like? What use would it be?

13.   What would be a good rule for how much to partition at any level; i.e., what is the maximum number of bubbles we should allow on any DFD?

If I have chosen an apt example, and if my set of conventions is as conceptually natural as I hope it is, you should have been able to answer all or most of the questions. The rest of this chapter should reaffirm your conclusions and offer some additional observations and examples. The answers to the questions above can be found at the end of this chapter.

7.2 Elements of a leveled DFD set

Not surprisingly, the leveled set of DFD's is made up of a top, a bottom, and a middle. The top is a single diagram called the *Context Diagram*. The bottom consists of a set of unpartitioned bubbles, called *functional primitives*. The middle is everything else.

7.2.1 The Context Diagram

In keeping with the philosophy of showing examples first, Fig. 32 represents a Context Diagram. It is, in fact, the Context Diagram for the set of DFD's shown in Fig. 31; it really ought to be thought of as part of the set.

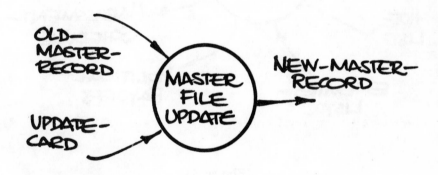

Figure 32

The Context Diagram is a "departitioned" version of the top-level breakdown of Diagram 0 — it shows only the net inputs and outputs. Drawing a Context Diagram may seem like a trivial formality. It serves only one purpose, but that is an important one: *to delineate the domain of our study.* Diagram 0 does this as well. I propose, however, that you take the extra five minutes to draw the (admittedly redundant) Context Diagram, as a formal declaration of the domain of study.

The Context Diagram documents the domain of study by showing the set of data flows that cross into and out of the domain. It is these net data flows that actually define the boundary. Usually the single bubble of a Context Diagram is given the name commonly ascribed to the area being studied. In most cases, the name will only be an approximation of the true domain. Look at the sample Context Diagram shown in Fig. 33. In spite of the name in the bubble, there is quite a bit more going on inside the domain shown than just "Municipal Taxation." The name should be thought of as no more than a handy monicker. Naturally, you should try to pick one that is not misleading. But the name is not intended to be a rigorous delineation of context — it is the set of input and output data flows that serves this purpose.

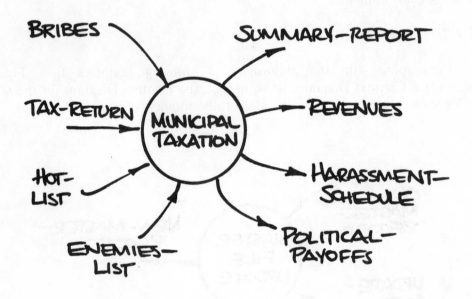

Figure 33

A Context Diagram has to be able to pass certain coherency tests. In order to apply these tests, think of the Context Diagram as a transformation, a process that transforms the input data flows into the outputs. Now ask the question, Is such a transformation possible? Is the set of inputs sufficient to build the output? If you find that you have the equivalent of apples and grapes coming in and orange marmalade going out, your Context Diagram is incoherent.

7.2.2 Functional Primitives

Functional primitives are bubbles which are not further decomposed into successively lower-level networks. Our decision to call something a primitive implies that it cannot be further decomposed (it has a kind of "atomic binding" that simply defies further partitioning), or that we have arbitrarily decided not to partition further because our requirement is satisfied. Figuring out which are the functional primitives of our system and how they interrelate is the goal of our Data Flow Diagramming effort.

Assuming that the leveled set of Fig. 31 is complete, all the bubbles shown in Diagrams 1, 2, and 3 are functional primitives.

7.2.3 The Middle Levels

If Fig. 32 is the top of our sample set of leveled DFD's, and Diagrams 1, 2, and 3 of Fig. 31 represent the bottom, then the middle must be Diagram 0. As is true of any middle-level figure, it portrays the breakdown of some or all of an area into a network of components which must themselves be broken down further. In general, there will be several middle levels, sometimes as many as eight or nine. In a very trivial example, we might have no middle at all. If we had chosen to portray the master file update as Fig. 32 plus Fig. 29 (note that these two together constitute a valid leveled set), then there would be no middle.

7.3 Leveling conventions

At the top of the hierarchy of a leveled set is the Context Diagram. It is the "parent" of the first-level breakdown diagram. That diagram is, in turn, parent to its child diagrams, which constitute level two. There are as many level-two diagrams as there are bubbles on the parent. Each level-two diagram may be the parent of some number of level-three diagrams. And so on.

7.3.1 The Parent-Child Relationship

Fig. 34 shows a parent figure together with one of its children. The parent in this example may have as many as five child figures; or it may have fewer than five, since one or more of the parent bubbles may be "primitive."

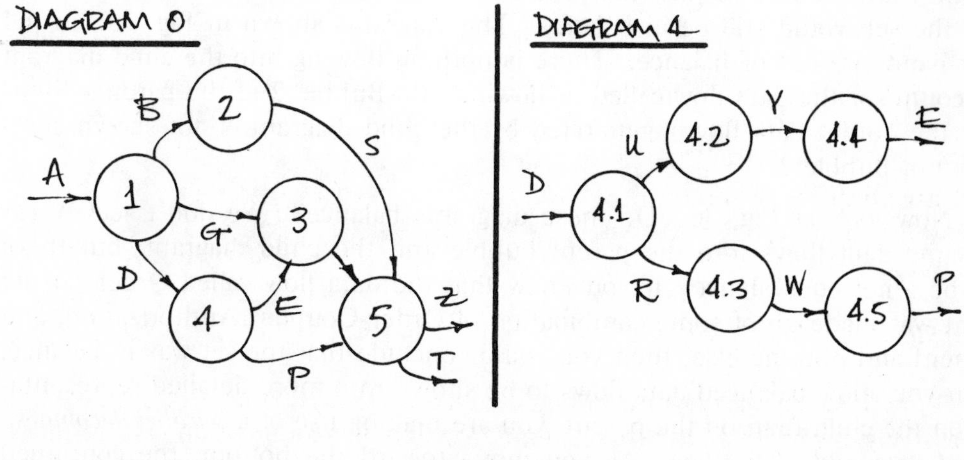

Figure 34

With explicit reference to Fig. 34, the relationship between the parent and child is this:

> The process represented by Bubble 4 on the parent is partitioned into the five component pieces shown on the child diagram. The child shows these pieces and the interfaces among them. It is a more detailed view of the transformation indicated on the parent diagram, in which an incoming flow of d's is transformed into e's and p's. Since the child presents exactly the same transformation as was portrayed by Bubble 4 on the parent, its net inputs and outputs are identical to the inputs and outputs to Bubble 4.

Diagram 4 (part of Fig. 34) might have as many as five children. One of these would be associated with Bubble 4.1. We would label it Diagram 4.1 to call attention to this relationship. As you must already have guessed, that figure would be another DFD showing a network of interconnected bubbles numbered 4.1.1, 4.1.2, and so forth. The net input to the diagram would be a data flow of d's and net output would be u's and r's. That DFD would be a further partitioning of the process shown by Bubble 4.1 in Diagram 4.

7.3.2 Balancing

Data flows into and out of a bubble on a parent diagram are equivalent to net inputs and outputs to and from a child diagram. This equivalence is called *balancing*. The balancing rule is as follows:

> All data flows shown entering a child diagram must be represented on the parent by the same data flow into the associated bubble. Outputs from the child diagram must be the same as outputs from the associated bubble on the parent with one exception: trivial rejects (reject paths that require no revision of state information) need not be balanced between parent and child.

Fig. 34 is balanced. Input and output data flows are the same for Diagram 4 as they are for Bubble 4 of the parent. If Bubble 4.3 had a trivial reject data flow, the set would still be in balance. The diagrams shown in Fig. 35, on the other hand, are out of balance. There is nothing flowing into the child diagram to account for the data flow called m flowing into Bubble 2 of the parent. Similarly, the output data flow s generated by the child diagram is not shown coming out of Bubble 2.

Now look at Fig. 36. Do these diagrams balance? We don't see *exactly* the same data flows into the parent bubble and the child diagram, but there may be a net equivalence. If you knew that the data flow called Order on the parent was made up of some combination of Order-Coupon, Authorization, and Payment and nothing else, then you could conclude that the set was in balance. When you allow balanced data flows to be shown in a more detailed representation on the child than on the parent, you are making use of a *parallel decomposition of data and of function*. As you move toward the bottom, the continued

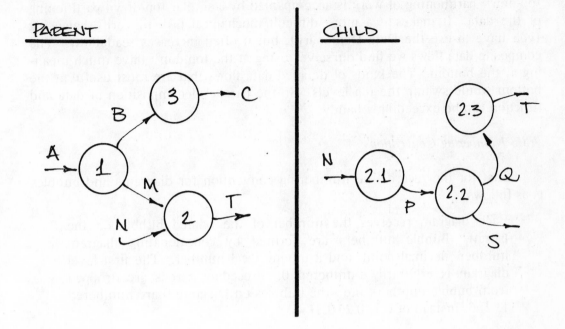

Figure 35

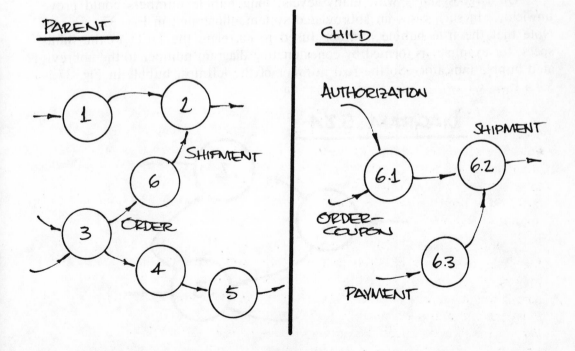

Figure 36

top-down partitioning of work is accompanied by a similar top-down partitioning of the data. It makes it a more difficult mechanical task to verify balancing (you have to use the Data Dictionary), but it often increases readability. The composite data flows we find ourselves using at the top don't have much meaning at the bottom. The kinds of detailed data flows that are most useful at the bottom would swamp the top levels. So the parallel decomposition of data and function can be exceedingly handy.

7.3.3 Numbering Conventions

Just for the record, the numbering convention for diagrams and bubbles is as follows:

Each diagram receives the number of the related bubble on the parent. Bubble numbers are formed by concatenating diagram number, decimal point, and a unique local number. The first level diagram is arbitrarily numbered 0. Preceding zeroes are dropped from bubble numbers (i.e., the bubbles on Diagram 0 are numbered 1, 2, 3, instead of 0.1, 0.2, 0.3).

Thus you can tell what level a given figure is by counting the decimal points in the numbers given to its bubbles. The level is always one more than the number of decimal points in the bubble numbers. It also follows that you can derive the number of the parent of a given diagram by removing the last digit of its diagram number; e.g., the parent of Diagram 6.3.2 is Diagram 6.3.

On large systems with many levels, long bubble numbers could prove unwieldy. In such cases an abbreviated system, illustrated in Fig. 37, is used. Note that the true bubble number (used to correlate the DFD to the mini-specs, for example) is formed by concatenating diagram number to the abbreviated bubble indicator. So the real number of the leftmost bubble in Fig. 37 is 5.8.4.1.

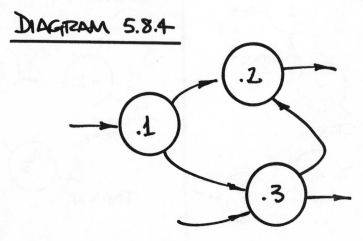

Figure 37

7.3.4 Local Files

Fig. 38 shows a parent and balanced child from a leveled set. Note that the file Alpha on the child does not appear at all on the parent. Before I explain, speculate on why that might be. You are bound to come up with the answer.

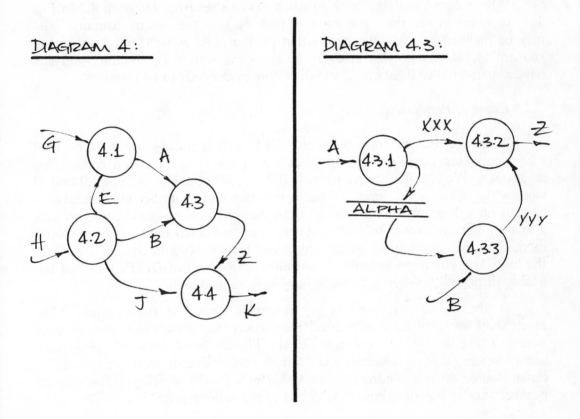

Figure 38

Alpha is entirely local to Diagram 4.3. It is not shown on the parent since it is not a meaningful interface there. For the level of abstraction shown in the parent diagram, it is important to suppress details relevant only to the insides of one bubble. So Alpha does not make an appearance in Diagram 4, for the same reason that the data flow xxx doesn't show up there.

The rule for when to show a file is:

A file is shown on a DFD at the first level where it is used as an interface between two processes.

A necessary consequence of using this convention is that: *at the first level where a file appears, all references to it are shown.* Taking the example of Fig. 38, this means that Alpha is not used by any processes in the entire set except for 4.3.1, 4.3.3, and their descendents. Since the direction of data flows is relevant, we would expect 4.3.1 and family only to write to the file, and 4.3.3 and its family to use it only for input.

7.3.5 Source and Destination Information

Where does the data flow z go when it vanishes from Diagram 4.3 in Fig. 38? In order to see that, you have to look back at the parent diagram. You may be inclined to mark the destination on the child as well, but I encourage you not to. If you do that at every level all the way to the bottom, you will have a maintenance headache if the data flow ever needs to be rerouted.

7.3.6 Extent of Partitioning

When we use any form of leveled DFD's, it is because we have decided that partitioning directly into primitives would result in too much partitioning all at once. We feel that diagrams with 100 or 1,000 bubbles are too difficult to work with. What is the limit? What is the maximum number of elements we can put on a diagram without obscuring its message? If we could come up with a nice numerical answer to that question, then we could limit ourselves to no more than that many processes at any given level. Accordingly, we could use the limit as a guideline to build a minimum leveled set of DFD's that had the quality throughout of being "conceptually easy to deal with."

To the extent that there is any simplistic answer to the question, How much shall we partition at any level? the answer is, *Partition into seven or fewer pieces.* There is a lot of evidence that the human brain deals efficiently with sets of seven or fewer elements, and much less efficiently with larger sets. A classic source on this syndrome is G.A. Miller's charming essay, "The magical number seven, plus or minus two," cited in the Bibliography.

If you are inclined to be dissatisfied with simplistic answers, you are probably right. I have certainly seen DFD's of only five or six bubbles that were terribly obscure, while some with a larger number of bubbles are nonetheless crystal clear. The final standard has got to be your own judgment on resulting readability. I offer you my observations on the subject:

- Listen to the data — if it cries out to be distributed in some fashion, allow it to affect your partitioning. A partitioning that is in this sense "natural" is often easier to understand than one that stays within any artificial partitioning limits.

- Partition to create conceptually meaningful interfaces.

- Expect the extent of partitioning (number of bubbles, for instance) to be somewhat greater at the very top level. The top-level diagram need not be readable at a glance, since it is a working document that you learn to live with.

- At any level, partition as much as you can without hurting readability. The more you partition, the fewer diagrams you will have to use and the fewer you will have to maintain in your Structured Specification.

- If you need an artificial limit, use seven — diagrams partitioned into approximately seven pieces are usually workable.

- Diagrams that are clear are better than diagrams that are not, regardless of "absolute" partitioning rules.

7.3.7 Summary of Leveling Conventions

The following is a summary of leveling conventions:

1. To see the detail of a given bubble, look at the *diagram* with the same number.

2. Inputs and outputs are balanced between parent and child — data flows into and out of the parent bubble are equivalent to data flows into and out of the child diagram.

3. At any given level, only files and data flows that are interfaces among DFD elements are shown. Files and data flows that are only relevant to the inside of some process are concealed.

4. At the first level where a file is shown, *all* references to it must be shown.

7.4 Bottom-level considerations

Each time we push our partitioning down another level, we divide our smallest segments into still smaller pieces. Obviously, this process has to stop someplace. The place where it stops is called (colorfully enough) the bottom. This gives rise to a number of questions:

- How do we decide where to stop partitioning — what constitutes an acceptable bottom?

- How do we mark the bottom level?

- Since bottom-level bubbles are not documented by further decomposition, how shall we describe their contents?

- How do bottom-level diagrams relate to each other?

The following subsections deal with these questions.

7.4.1 Determination of the Bottom Level

I know of three different criteria for deciding when a partitioning effort has gone far enough. I'll state them all, and you can choose among them:

1. My own approach is to stop when I believe the insides of my lowest-level bubbles can be completely described in a mini-spec of about one page. This is obviously an arbitrary rule. It has, however, a nice side effect on readability. Mini-specs that are much shorter than a page are too choppy; and larger ones seem to me to cry out for further partitioning.

2. Some analysts try to continue partitioning down to the point where the bubbles have a single input data flow and a single output data flow. In applying this rule, you do not count trivial error paths. You may also have to accept a small number of unpartitionable bubbles that have more than a single entry and single exit. (You can quickly prove to yourself that in a merge, for instance, no matter how much you partition there will always be at least one bubble that has more inputs than outputs.)

3. Students and readers of Michael Jackson's work, *Principles of Program Design* (see Bibliography), have suggested that his concept of boundary clashes may be applicable to determination of the bottom level. Rather than try to treat the subject of boundary clashes here, I refer you to Chapter 7 of Jackson's book. Application of the idea to data flowing would involve searching for numerical relationships between input and output quantities. Clean one-to-one or many-to-one relationships imply that partitioning has gone far enough. If there is no such relationship — if, for instance, a bubble must transform three x's into two y's — then further partitioning is required.

Each of these three approaches will be useful at one time or another in your Data Flow Diagramming. I often find that trying to apply the first of these is an iterative business; you think of some bubble as primitive until you try to specify it — then you find yourself with an unwieldy description, so you go back and partition it further.

7.4.2 Marking the Bottom Level

How can a reader of your DFD's tell by looking at a given diagram that one or more of the bubbles on it is primitive? Is there some way to mark primitive bubbles? I advocate that you not mark primitive bubbles in any way; let your reader determine that a bubble is primitive by looking for, and not finding, a further decomposition of it. In other words, if you can't find Diagram 4.3.6, then Bubble 4.3.6 on Diagram 4.3 must be a primitive.

Students in my seminars often boo and hiss and throw things when I suggest this, so if the idea does not appeal to you, you are not alone. My reason for suggesting that primitive bubbles not be marked in the DFD is that the set of diagrams should be considered a working document — it should be expected to change. Frequently, what used to be called a primitive later needs to be decomposed, so it pays to make it easy to incorporate such a change.

Some analysts have adopted a slightly different approach: they mix their mini-specs in with the leveled DFD's. That way, you always find something when you drop down a level. If you find a mini-spec, that means the bubble was primitive. Note that this approach also has the advantage of being easy to change if you decide to partition further.

7.4.3 Documentation of Process

Look back at Fig. 38. Assume for the moment that each of the bubbles on Diagram 4.3 is primitive. That means that if you search through the leveled set, you are not going to find any lower-level diagrams in the 4.3 family. Since there is no further DFD description of Bubbles 4.3.1, 4.3.2, or 4.3.3, you have to document them in some fashion other than DFD description. You have to write a narrative or Structured English description or a Decision Table or perhaps even (Heaven Forfend!) a flowchart. In any event, that description you write to document the primitive is what I have been referring to as a mini-spec.

The rule for correlating mini-specs to the leveled DFD set is:

There must be one mini-spec for each DFD bubble which is not further decomposed, i.e., for each primitive. Mini-specs should be marked with the bubble number of the related bubble.

But how about the upper levels? How about Bubble 4.3 on the parent diagram shown in Fig. 38? It's not a primitive, but aren't we going to write a mini-spec for it anyway? The answer is no. If we did write a mini-spec for it, that document would be *100 percent redundant*. Once we have completed Diagram 4.3 and the three mini-specs describing its three bubbles, then Bubble 4.3 on the parent is already completely specified. Bubble 4.3 on the parent is *rigorously equivalent* to the network of Diagram 4.3. So when all those processes are described (via mini-specs) and the interfaces are defined (in the Data Dictionary), there is no further non-redundant specification possible. A necessary consequence of this line of reasoning is the following:

The number of mini-specs in a complete Structured Specification will be exactly equal to the number of primitive bubbles in the leveled DFD set.

7.4.4 Building Expanded DFD's

You might observe that the higher-level bubbles are only bookkeeping entities, ways to keep track of connected sets of lower-level bubbles. A demonstration that the upper-level bubbles are no more than sets of lower-level entities is the following: Working on Fig. 38, replace Bubble 4.3 on the parent diagram with the entire child network. Note that all the inputs and outputs match up perfectly. (If they didn't, the leveled set would have been demonstrably out of balance.) If you do the same thing for each of the bubbles on Diagram 4 (replace Bubble 4.1 with Diagram 4.1, Bubble 4.2 with Diagram 4.2, and so forth), you will come up with the expanded version shown in Fig. 39. Note that you have not changed the specification by doing this; you have simply presented it in a different format. The "system" is still the same set of primitives, interconnected in exactly the same manner. Only your portrayal of them has changed.

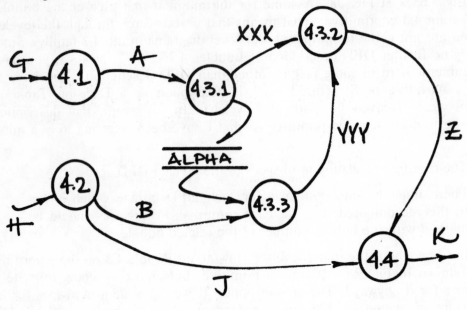

Figure 39

If you took the entire expanded network of Diagram 4 in the figure and used it to replace Bubble 4 of its parent (its parent would be Diagram 0, not shown), you would be building a still more expanded variant of your system. In fact, if you replaced every non-primitive bubble by its lower-level network, and you did this in a systematic bottom-up fashion, you could build a super-expanded network representing the whole system. All of the bubbles on that diagram would be primitives. That expanded network of primitives would be the football field of paper that we introduced the idea of leveling to avoid.

I have presented the concept of expanded versions because it is important for you to understand that such a thing is possible and why it is possible. We have no more interest now than we ever did in the football-field-sized diagram. But expansions will nonetheless turn out to be useful — we just won't carry the idea to its extreme.

7.5 Advantages of leveled Data Flow Diagrams

I hope the advantages of a leveling approach are more or less obvious to you by this time. They are listed below:

- Leveled Data Flow Diagrams allow a top-down approach to analysis. In particular, they help us build a specification which can be read top-down. That means a manager can restrict his reading to the top few levels and still get the big picture. Implementors and users can read from the abstract to the detailed, narrowing in on particular areas of interest.

- There are no true off-page connectors. Data flows vanish off the page (go back up into the context of the parent), but each page is a complete presentation of the area of work allocated to it — you never have to find another page and continue reading there to get the whole description.

- All diagrams can be conveniently restricted to 8 1/2 x 11 inch paper (or A4 if you are metric). That means that the entire Structured Specification can be copied using conventional photocopiers.

7.6 Answers to the leveled DFD Guessing Game

I asked you earlier to try to guess in advance how leveled DFD's ought to be expected to work. I hope the idea is natural enough that you were indeed able to predict most of what I was going to say. In case you stumbled on any of the Guessing Game questions, let's go back and fill in the answers. All diagrams referred to are shown in Fig. 31.

1. Which diagram might be thought of as the "parent" of Diagrams 1, 2, and 3? (Diagram 0)

2. What is there on Diagram 0 that corresponds to the net inputs and outputs to and from Diagram 2? (The data flows into and out of Bubble 2.)

3. How many levels are shown here? (Two)

4.   Why do you suppose I stopped my partitioning where I did? (It was pointless to go on — the pieces were small enough to be described completely in a single page of text; most of them were single input, single output, and there were no boundary clashes.)

5.   If I did manage to come up with a further partitioning of Bubble 1.2, for instance, and I broke it up into three pieces, what would I number the diagram that showed those three pieces? (Diagram 1.2)

6.   What would be the net inputs and outputs to and from that diagram? (The same as the input and output data flows of Bubble 1.2, i.e., Sequence-Checked-Card and Edited-Card.)

7.   What numbers would the bubbles on that diagram have? (1.2.1, 1.2.2, and 1.2.3)

8.   How many levels would there be then? (Three)

9.   Should I be disturbed that my partitioning dropped down into more levels in one area, and fewer in another? (No) What does that imply? (That my partitioning was not absolutely even — some of the pieces were bigger than others, and so required more levels of subsequent partitioning.)

10.   Have I really specified anything with Fig. 31 alone? (No — I have only partitioned.) What more is required to complete the specification process? (A mini-spec for each primitive.)

11.   Suppose that Fig. 31 is a complete leveled set of DFD's, i.e., no further decomposition is required. How many mini-specs will I have to write to specify all the processes non-redundantly? (Nine) What is the rule for correlation between the set of mini-specs and the leveled DFD's? (The number of mini-specs is equal to the number of primitives. Bubble numbers of the primitives ought to be picked up on corresponding mini-specs.)

12.   Can you imagine another level "above" the level shown in Diagram 0? (Yes, the Context Diagram.) What would it look like? (Fig. 32) What use would it be? (Formal declaration of the domain of study.)

13.   What would be a good rule for how much to partition at any level; i.e., what is the maximum number of bubbles we should allow on any DFD? (Just not too many to obscure readability. Six or seven for most diagrams.)

8 A CASE STUDY IN STRUCTURED ANALYSIS

It's time for an example. This chapter will use the tools of Structured Analysis to show you the workings of a small enterprise. Assume that you are a new analyst, just coming on board a project to automate some of the company's procedures. It is still early in the analysis phase, so all that has been completed is the study of current operations — we don't even know yet what is to be automated.

The case study material includes a partial Data Dictionary. I haven't yet said much about the use of a Data Dictionary in Structured Analysis. And, in particular, I have not said anything about our convention for writing definitions. But I am going to ask you to look at and try to understand a number of Data Dictionary entries associated with the case study. Don't worry about the convention, which may seem foreign at first. By the time you arrive at the Data Dictionary chapters, I predict that you will already have picked up the most important concepts.

The Data Flow Diagrams you are going to see, together with the supporting Data Dictionary, represent what I have called the current logical model.

8.1 Background for the case study

During the 1960's in the United Kingdom, a number of Great Systems Thinkers and Captains of Industry got together and formed a company whose charter was to extract money from that good-sized segment of the English public which is addicted to horse racing. The idea behind the company was simplicity itself: People who bet on the races are inclined to be superstitious, so a service that used astrology to predict race results might be very saleable. The company that was born was called the Astro-Pony Toutshops, Ltd. (APT).

The service provided guidance to clients on when to bet. As a function of dates and times of birth sent in by customers, the company would prepare astrological profiles to tell them when to risk their money and when not to. The profiles were correlated to the race schedule. No advice was given on which horse to pick. The theory was that "when you're hot, you can't miss."

APT started off as a mail-order operation. Customers filled in coupons cut out from an advertisement or reorder coupons supplied with previous profiles. After a year or so, a new gimmick was added that incorporated information about the horses' dates and times of birth as well. It was a big hit —

evidently people who believed their own fortunes were controlled by the stars and planets were easily convinced that the same thing applied to horses. At this point, the operation was still totally manual. But in those Salad Days of EDP, could computerization be far behind?

8.2 Welcome to the project — context of analysis

Fig. 40 is the Context Diagram for the Astro-Pony Toutshops operation. It defines the domain of our study. As indicated by this diagram, we will limit our analysis to those portions of the company that accept Orders and Mail-Payments from Customers and send back Profiles or Delinquency-Notices to them, plus those portions that effect the interfaces shown to the Stables and Tracks (which seem to provide data base information), the Bank, the Salesmen, and a Credit Card Company.

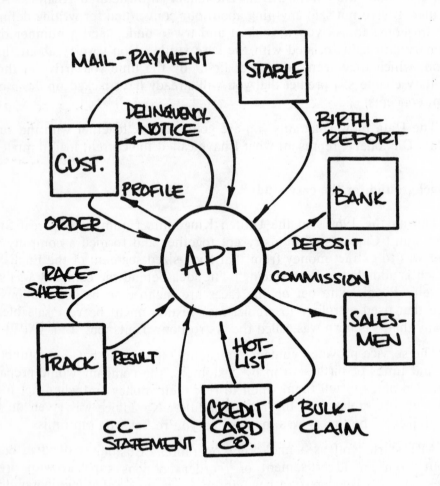

Figure 40

It is the data flows into and out of the Context Diagram that delineate the boundary. The name given to the diagram is only an approximation. In our case, the true domain of study is only a portion of Astro-Pony Toutshops, Ltd. For instance, you might expect the company to generate and send out advertising, but that is clearly outside of the context. It has not been included as part of the study because the analysts decided it was not relevant.

Let's apply some coherency tests before going further. The Context Diagram shows money and checks flowing out in the form of a Deposit going to the bank, so there had better be some depositable items flowing in. The Profile that goes out to the customer presumably has to have his name and address on it, so that information has to flow in from someplace. Are there any obvious failures — outputs that simply could not be generated using only the inputs shown? In order to see this, you have to refer to the Data Dictionary where the composition of the various data flows is recorded. For instance, the Data Dictionary entry for Order

> **Order** = **Customer-Name-and-Address +**
> **Time-and-Date-of-Birth +**
> **(Race-Specifier) +**
> **(Payment)**

tells us that Orders are made up of four components, the last two of which are optional. The first component is Customer-Name-and-Address, so that information required to build and send out a complete Profile is available. At least sometimes there is a Payment, included as part of an Order. That means that there is some inflow of depositable items required to make up a Deposit. So far, there is no demonstrable incoherency.

8.3 The top level

Fig. 41 is the top-level partitioning of our context area. As is often the case with descriptions of the current environment, this one shows partitioning more or less along organizational lines. This may not be the most functional choice, as I have defined the word, but it is usually a good start. It has the great advantage of being familiar to the user, so he can give us all the more assistance in making sure it is right.

What follows is a walkthrough of Fig. 41 and the rest of the leveled DFD set, conducted along the lines of an analysis phase walkthrough for a user. It begins with a quick overview of the top level, preparatory to plunging deeper into one of the detail areas.

The most important incoming data flows are the Order and the Mail-Payment, both of which originate with the customer. The Order is handled first by the Sales Department. Some Orders arrive through the mail, while others are phoned in to a salesman. For the purposes of a DFD, the two are the same (although only mail Orders can include payment by check). Sales processes the incoming Order and redrafts it into the form of an Invoice:

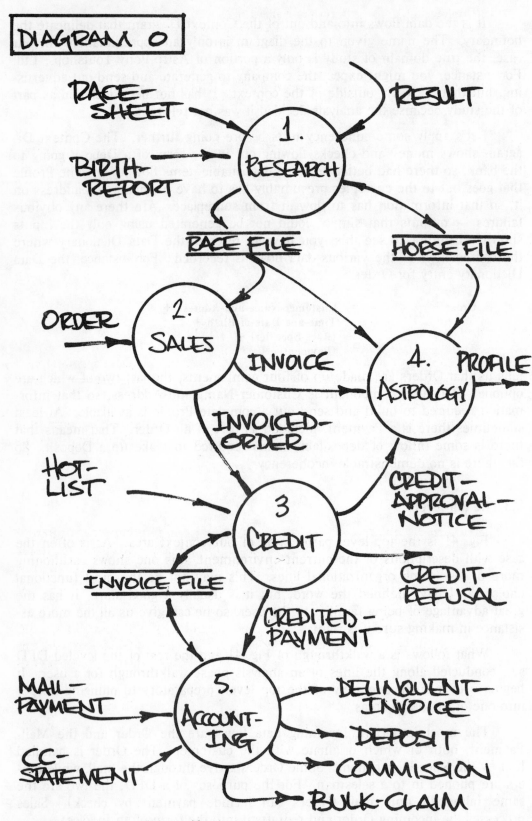

Figure 41

```
Invoice              =   Invoice-Number +
                         Customer-Name-and-Address +
                         Time-and-Date-of-Birth +
                         (Race-Specifier) +
                         Price +
                         (Amount-of-Payment + Mode-of-Payment) +
                         (Salesman-ID) +
                         Net-Amount-Due
```

The invoice is sent to the Astrology group, which adds the actual predictive information to the Invoice and sends the whole package out, regarding it as a Profile:

```
Profile              =   Invoice +
                         Custom-Prediction +
                         Super-Personalized-Astrological-Toutsheet +
                         Reorder-Coupon +
                         (Voucher-Copy)
```

A pink copy of the Invoice, together with any payment that arrived with the Order, is sent down to the Credit Department. (The fact that it is a pink copy is a detail not contained anywhere on the DFD or the Data Dictionary. I often add such information during the walkthrough to help the user relate back to the environment as he knows it.)

An element of both the incoming Order and the Invoice is something called a Race-Specifier, optional in both cases. Our Data Dictionary tells us that:

```
Race-Specifier       =   Race-Date + (Race-Number) + Track-Name
```

The fact that it is an optional item implies that there are really two different services provided by APT: It gives you a Profile for a given day at the track (possibly limited to a specific race), or it picks some situation which is "in your stars" for the near future. Prices, by the way, depend on which service is performed.

On credit sales, the customer's past payment history is scanned by Credit, causing a Credit-Approval-Notice to be sent to Astrology. Obviously, no unpaid item is going to be sent by that department until the Credit-Approval-Notice catches up with the Invoice.

If all goes well, the Invoice and the Credited-Payment (if any) are passed on to Accounting. The Payment that comes into Credit as a part of the Invoiced-Order is either a Check or a Credit-Card-ID:

```
Payment              =   [Check | Credit-Card-ID]
```

The method of payment affects how Credit does its job, as does whether or not payment was included. If payment was by credit card, a check is performed on the hot list of bad cards, and a voucher is made up. One copy of the voucher goes back to Astrology as part of the Credit-Approval-Notice:

> **Credit-Approval-Notice**   =   **Invoice-Number + (Voucher-Copy)**

where it makes its way into the package called Profile, and thus gets sent to the Customer. The other copy of the voucher goes to Accounting as a Credited-Payment:

> **Credited-Payment**   = |Check | Voucher|

The Accounting Department deals chiefly with funds that arrive as components of certain incoming data flows (Credited-Payments, Mail-Payments, or CC-Statements) and with information about expected funds (orders that have no payment or only partial payment attached). Accounting is the major interface to the Credit Card Company. When vouchers arrive from the Credit Department, Accounting groups them together and sends them out to the Credit Card Company in the form of a Bulk-Claim. The term Bulk-Claim refers to a special Claim-Form with the company's name and address and ID number on it, along with however many vouchers have been written up:

> **Bulk-Claim**   =   **Claim-Form + {Voucher}**

When checks arrive from any source, Accounting deposits them. Mail-Payments require that information be sent back (via the Invoice File) to the Credit Department. Accounting also pays Commissions to the Salesmen. On credit accounts, the commissions are paid only when and if the actual Mail-Payment is received.

The function of Research is rather obvious: to build the Race-File and the Horse-File for use by the other departments. The Race-File correlates horses to track and race number:

> **Race-File**   =   {Track-Name +
>   Race-Number +
>   Horse-Name}

and the Horse-File has astrological data about each animal.

8.4 Intermezzo: What's going on here?

Before we proceed to the next level, it is worthwhile to look back and see what has happened so far. *How much of what I have just told you was really in the DFD and the Data Dictionary, and how much was ad-libbed?* The main thing I added during the walkthrough (that would not have been apparent had you been reading the diagram without my help) was procedural information. I alerted you to the fact that Orders into Bubble 1 cause the creation of Invoices *and* Invoiced-Orders, for instance, and that Invoiced-Orders arriving at Bubble 3 are transformed into Credit-Approval-Notices *or* Credit-Refusals, and that in the case where the Credit-Approval-Notice is sent out, an Invoice is sent to the Invoice-File *and optionally* a Credited-Payment is sent to Accounting.

Without someone to present it, such procedural information is just not available to the reader of a high-level DFD. If you react as I did when I first started to work with Data Flow Diagrams, you probably have two serious reservations on this point:

- Where do you have to go in the Structured Specification to find the procedural information?

- How useful is a Data Flow Diagram if it doesn't show procedural flow? Does it mean that nobody can ever work alone on a DFD, that each and every user and each new analyst and implementor will have to be spoon-fed and walked through each diagram?

All the information I ad-libbed comes from the lower levels; much of it comes from the very bottom. In particular, the correlation between input and output data flows is not on any of the DFD's — you have to read all the way to the bottom and then look in the referenced mini-spec to get it. In an eight-level set, it might involve substantial effort to ferret out the procedural connections. But they are there, documented in the mini-specs.

As for the usefulness of the diagrams without the procedural connections readily at hand, I honestly don't think it is affected. True, it may be a struggle (at least at the top few levels) to follow a particular item through the system, but describing data movement was never the purpose of a DFD. Its purpose was to *divide the system into pieces* in such a way that the pieces were reasonably independent, and to *declare the interfaces among the pieces.*

I suggest that trying to understand the system by following items serially through the DFD is using it improperly. It is this serial means of absorbing the subject matter that has hampered us in the past. With a DFD, you have the opportunity to absorb the information in parallel; not to *read it,* but to *see it.* If the DFD is skillfully conceived and meaningfully labeled, your grasp of its message should be complete enough after a minute's study of the diagram to answer questions such as: If there is any (whatever) being done in this system, in which segment is it happening? And though you may not yet be sure of specifically what is done and how it is done, you know precisely where to look for everything.

Now go back to Fig. 41 and try to *see* it. The picture there is a static one; no movement is readily apparent. What is shown is a whole, divided up into five pieces. At a glance, you should have a feel for what is going on and some idea of how the segments relate. Your understanding of the system is not complete, but your understanding of the partitioning is. The picture that is communicated is not a procedural one, but in spite of that — or because of it — it is *useful.* Viewed in this fashion, the true nature of the DFD is apparent: It is not a flow diagram in the traditional sense, but a tool for functional decomposition. Once you learn to use it in this manner, you will need no guidance to understand its content. You are still going to have to do some walking through to help users and new project personnel get started. But the purposes of these walkthroughs will not be so much to help them understand the subject matter, as to make them conversant with the methods you have used. Your principal goal in such presentations is to make your listeners self-reliant.

If you tried to verify balancing between Fig. 41 and its parent (the Context Diagram presented in the previous figure), you may have noticed a problem. The parent shows something called a Delinquency-Notice that does not appear on the child. The child diagram, on the other hand, has two outgoing data flows — called Delinquent-Invoice and Credit-Refusal — that do not appear on the parent. Are the two diagrams therefore out of balance? No, this is one of those cases where you need to refer to the Data Dictionary to verify balancing. You have to look there to determine that the thing called Delinquency-Notice on the Context Diagram is made up of either a Delinquent-Invoice or a Credit-Refusal:

Delinquency-Notice   =   [Delinquent-Invoice | Credit-Refusal]

There is another problem evident in Fig. 41, one that cannot be explained away. It has to do with Bubble 5. Think of Bubble 5 and its input and output data flows as a little Context Diagram for its child, Diagram 5. Exactly the same rule applies to an upper-level bubble as applies to a Context Diagram — it has to be coherent. Data cannot flow out unless it flows in in some form. I suggest that there is something intrinsically unworkable about Bubble 5 the way it is shown. It has to do with computing the Commissions. See if you can apply a test of reasonability and detect the problem. I'll cover it at the next lower level in the walkthrough of Diagram 5.

8.5 The lower levels

Fig. 42 and Figs. 44 through 47 present the lower-level diagrams for our case study. In everything that follows, I shall refer to them in terms of the diagram numbers rather than the figure numbers, since the diagram numbers have some significance to us (they show how the diagrams interrelate) and the others do not.

Note that the bubble and diagram numbers do not indicate any form of precedence. There is nothing magical about Bubble 1, just because it has the first number. It is not necessarily the first one you ought to consider. Just to hammer home this point and to break down your residual inclination to look at the diagrams procedurally, I am going to begin the second-level walkthroughs with Diagram 4.

8.5.1 Profile Generation

Look first at Bubble 4 of the parent Diagram 0 and imagine Diagram 4 to be what you see when you zoom in on the bubble. Whatever we see on the child diagram is going to have to portray the transformation of Invoices and Credit-Approval-Notices into Profiles shown on the parent. In Diagram 4, we see that transformation divided up into six component pieces, and we see the interfaces among the pieces.

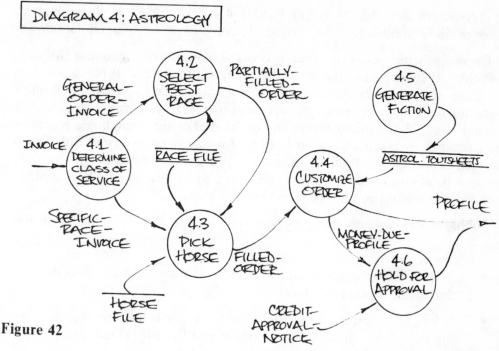

Figure 42

Bubble 4.4 is where the Filled-Order has the company's custom product, the Super-Personalized Astrological Toutsheet, added to it to make up the Profile. The toutsheets are taken out of a library file of stock gibberish. The basis for selecting which sheet to include must be something contained in Filled-Order, since that is the sum total of incoming information available to the process. Our Data Dictionary tells us that the Filled-Order is made up of:

> **Filled-Order** = **Invoice +**
> **Race-Specifier +**
> **Horse-Name + Horse-Astrol-Sign +**
> **Customer-Astrol-Sign**

Presumably, it is the combination of astrological signs that is used to pick a particular Super-Personalized Astrological Toutsheet. Since there is no shared data aside from the data flow between Bubbles 4.3 and 4.4, you have to conclude that there must be some Super-Personalized Astrological Toutsheet in the file to justify any imaginable horse that is picked. Of course that is exactly the way it works; what else would you have expected?

Bubble 4.5 is something you will not see terribly often, a process that produces output but has no input. If you find such a thing on DFD's that you are building, you ought to go back to the user and see where you have misunderstood his operation. It is a sure sign of missed communication. Either that or a sign of fraud, as in the case of APT. If you ever do come up with such a situation, and it is a correct representation of the operation, then there is no need to struggle over a name for the bubble; use the one I have provided for Bubble 4.5, a good general-purpose name for output-only processes.

Profiles are sent out by either Bubble 4.4 or Bubble 4.6, depending on whether or not any amount is due. Note the convention for merged data flows.

Common sense tells you that Process 4.6 has more incoming Money-Due-Profiles than outputs. Some of the completed invoiced Profiles seem to die in there. In this particular case, there is a reasonable explanation: Profiles that are mismatched (for which no approval arrives) are thrown away at the end of the day. The required processing of refused credit, including a response to the customer, is taken care of in Credit. All that is necessary in Bubble 4.6 is the destruction of the Profile. Whenever you detect a situation where data comes to a dead end inside a process, you have to be suspicious. It is another sign of possible missed communication with the user.

Bubble 4.3 is certainly going to be a functional primitive — it has all the signs:

- The valid name given to it is made up of a strong action verb and a single concrete object.

- It has single input and single output. Really it has two incoming data flows, Specific-Race-Invoice and Partially-Filled-Order, but from its point of view, the two are identical.

- There is no structure clash. Each input will eventually cause one output.

- Its allocated work can easily be described in one page or less.

As an indication of this last point, I have prepared a Structured English description of Process 4.3 and included it as Fig. 43.

8.5.2 Sales

The Sales Department is next. As you can see from Bubble 2 of Diagram 0, Sales has to effect the transformation of incoming Orders into Invoices and Invoiced-Orders. Go back to the definitions for these items if you have forgotten their composition. To find the details of Bubble 2, we go to Diagram 2.

Here we see some validity checking going on. The first check is dependent on no more than the content of the Order itself, whether it has been filled out properly. The second check is performed only on orders for a specific race

PROCESS NUMBER: 4.3

PROCESS NAME: Pick Horse

For each arriving order (Partially-Filled-Order or Specific-Race Invoice), do all the following:

    Pull Race-File folder for the race.

    Look up Julian day number of Customer-Birth-Date.

    For each horse in the Race-File folder:

        Calculate difference in days between Horse-Birth-Date and Customer-Birth-Date. Express difference as a positive number.

    Select horse(s) for which difference is closest to 23.

    If more than one selected,

        Order alphabetically by Horse-Name.
        Select last one.

    Make up Prediction card with Horse-Name and Horse-Astrol-Sign.

Figure 43

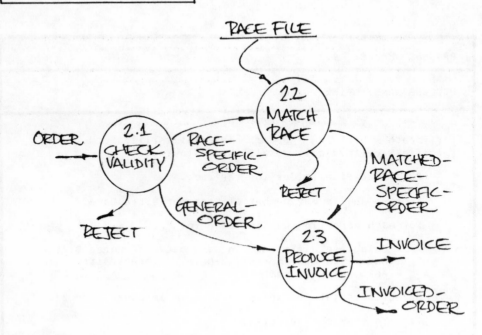

Figure 44

or day at the track. The salesman has to make sure that there is sufficient information in the Race-File to fill the order. In either case, the reject is considered "trivial," as I have defined the term, since there is nothing to be undone. The rejected order is sent back to the customer. In keeping with the convention, reject paths are not pursued further and are not balanced upward.

Once verified, the information on the order is used to create an Invoice (Bubble 2.3). The Invoice is sent on to Astrology. You actually have to look back at the parent diagram to see where it goes. The Invoiced-Order (Invoice copy plus Payment, if any) is sent down to Credit.

There is definitely some information coming out of Bubble 2.3 that never went in. If you look back at the definition of Invoice, you will see that it has the price of the service on it and the Salesman-ID. Neither of these items is contained in the inputs to the process. The idea of a process being a net source of information of any kind has to make you very suspicious. You have to be sure that there is not a hidden (undeclared) interface to some other process. If there is, then our DFD is wrong and must be corrected. If the originating data is of a constant nature and not used by any other process, then it is acceptable. This applies to Bubble 2.3.

8.5.3 Accounting

Accounting is shown on Diagram 5. As we already knew from the parent, Accounting has to transform three different kinds of payment input (Mail-Payments, Credited-Payments from Bubble 3, and CC-Statements from the Credit Card Company) into Deposits, Commissions, and Bulk-Claims for future CC-Statements. It must also put out Delinquent-Invoices under some conditions. Diagram 5 presents the further partitioning of this work.

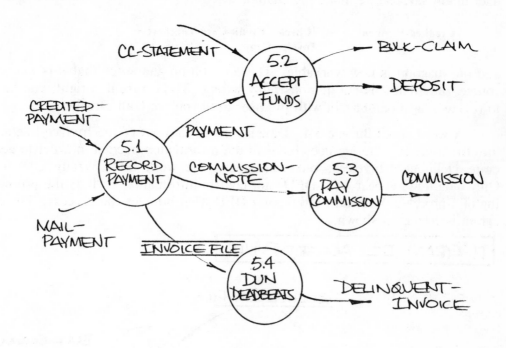

Figure 45

I stated earlier that there was a coherency problem with the work allocated to Diagram 5. See if you can spot it here. The thing called Commission-Note must obviously have information about which salesman was responsible for the sale and either the amount of the commission or of the sale itself:

$$\text{Commission-Note} = \text{Salesman-ID} + \text{Amount-of-Sale} + \text{Invoice-Number}$$

But how can Bubble 5.1 possibly get this information in order to make the Commission-Note? Mail-Payments may contain an invoice copy or at least a customer name, thus allowing the original invoice to be found; and the invoice has the Salesman-ID on it. So it works all right for that kind of input. The Credited-Payments, on the other hand, have no information on them about

Salesman-ID or any invoice number to retrieve the original invoice:

Credited-Payment   = |Check | Credit-Card-Voucher|

So in the case of direct payment, Accounting cannot possibly make up a Commission-Note.

The diagram does not pass its consistency check. Bubble 5.1 has to be a net generator of Salesman-ID's and Invoice-Numbers which is unreasonable. Since it can't possibly work the way it is shown, the analyst has to go back to the user and find out how it does work.

"The payment is always accompanied by the yellow invoice copy," the user might say. So we make the change

Credited-Payment   = |Check | Credit-Card-Voucher| +
Invoice-Copy

and the diagram is now workable. There is still no guarantee that it is *right*, of course, only that it is not demonstrably wrong. To be sure it is right, you have to review it again and again with the user until you are both satisfied.

A word about Bubble 5.4. Data seems to leap out of the Invoice-File and into the process. The file does have all the information needed in order to generate Delinquent-Invoices, but what kicks the work off? Wrong question. Only data flows appear on a DFD, not control information such as the prompting of a process. You can't tell from a DFD what initiates the process. The diagram is correct as shown.

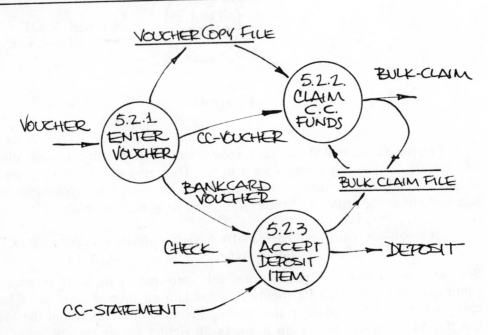

Figure 46

Look at Bubble 5.2, the transformation of Payments and CC-Statements into Bulk-Claims and Deposits. Where would you look to see the details of that transformation? If the bubble is primitive, there ought to be a mini-spec labeled Process 5.2, and if it is not primitive, there should be a Diagram 5.2. In this case there is a diagram (Fig. 46) showing the further decomposition of function. Diagram 5.2 is a third-level DFD. Let's look there next.

Diagram 5.2 is in balance, but you need information from the Data Dictionary to see that. You'll find there that Payment is made up of either Voucher or Check. This is one of those cases where the analyst has chosen to portray a parallel decomposition of data as he moved down a level in the DFD's.

I'll leave you to work out the rest of Diagram 5.2. In particular, speculate on why Bubble 5.2.2 needs the two files to which it has access. There is also something interesting shown here about the different ways companies are obliged to deal with credit card vouchers.

8.5.4 Research

Diagram 1 presents the work of the Research Department: preparation of the Race-File and Horse-File for use of the other areas. It ought to speak pretty much for itself.

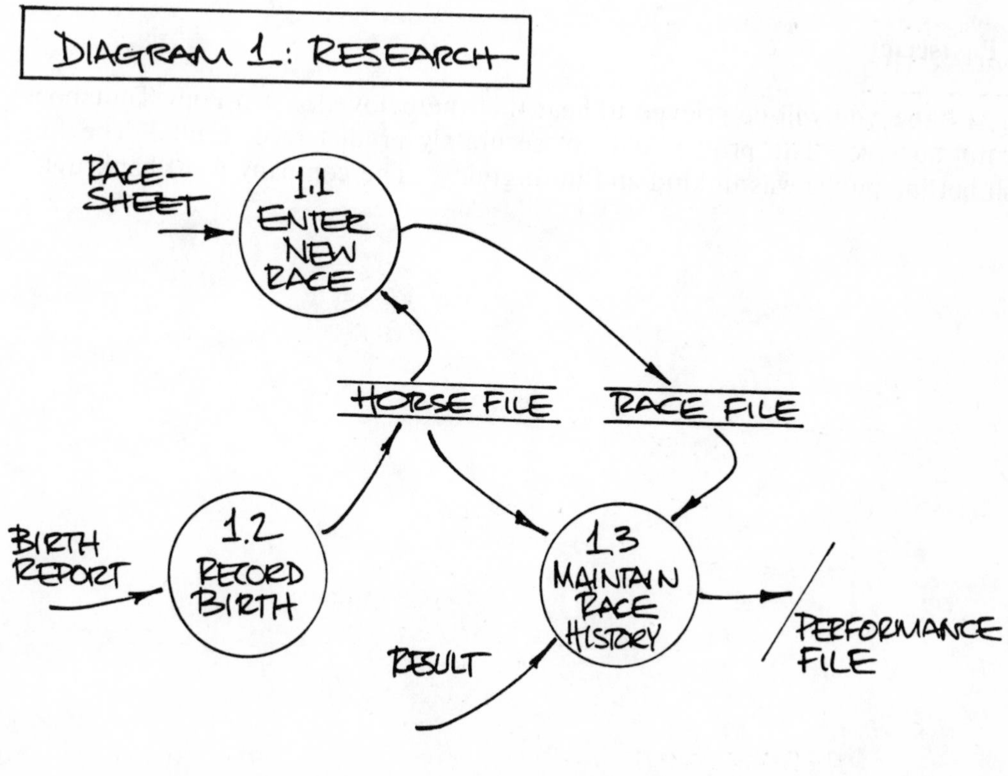

Figure 47

Notice that results of prior races are not incorporated into the Horse-File but go instead into the Performance-File. The Performance-File is local to Diagram 1; you know that because it did not appear on Diagram 0, the parent. But the first time that a local file shows up on a DFD, we ought to see *all* references to it. In the case of Diagram 1, that means that information is stored there by Bubble 1.3, but then never used by anyone else. Usually, the occurrence of a "write only" file on a DFD means that you have misunderstood your user. You must go back and ask him what the file is intended for. As shown, it is probably an error. I have included it in Diagram 1 as an example of the kind of inconsistency that often turns up on DFD's when you first draw them.

8.6 Summary

My reason for presenting the case study material here was a double one: First, to give you an example of the use of the tools of Structured Analysis; and second, to provide a vehicle for you to try out the tools and techniques yourself. I have shown you all the second-level diagrams except for Diagram 3. Try to draw a feasible Diagram 3 that balances with the rest of the set. There is no "right" answer — your DFD will reflect your concept of how the Credit Department probably works. But make sure it is not demonstrably wrong.

8.7 Postscript

I know you will be grieved to hear that the beloved Astro-Pony Toutshops did not survive. The profiles did not accurately predict race results. The English betting public was unkind and unforgiving. The company went bankrupt.

9 EVALUATION AND REFINEMENT OF DATA FLOW DIAGRAMS

It is all very well to say that the human mind works iteratively, and that we should therefore go about analysis, and in particular our Data Flow Diagramming, in an iterative fashion. But what is the basis for each iteration? How shall we guide ourselves to detect the failings of a tentative DFD, and to come up with a better version?

The purpose of this chapter is to document the set of techniques used to prove and improve Data Flow Diagrams. These techniques fall into the following three categories:

- *Tests for correctness,* sure signs that a DFD is wrong, that we have misunderstood the user or failed to translate his description of operations into a structured format properly.

- *Tests for usefulness,* indications that a DFD, although it might be technically correct, is over-complicated and difficult to understand, that it does not live up to our requirement of a conceptually easy-to-deal-with description.

- *Approaches to starting over,* mechanical methods for coming up with an improved version.

9.1 Tests for correctness

Usually, the first few times you try to draw a DFD that describes a given situation, you end up with a result that is just plain wrong. It is simplistic, demonstrably incorrect, unworkable, doesn't take account of known features, and doesn't hang together. So why did you even bother to write it down on paper? The answer is that you know the value of an early version, no matter what its failings. You know that if you wait for a complete and perfect concept to germinate in your mind, you are likely to wait forever. Perfect ideas do not germinate, they evolve. So you put your lousy idea down on paper, rout out its faults one by one, and gradually come up with a good product. There is a name for this careful improvement of a faulted concept into a desired result: It is called engineering.

Engineering of a Data Flow Diagram begins with the application of a set of tests for correctness. What would be handy for this is a list of the ways in which a DFD can be wrong; here is mine.

- It might have missing data flows, information required by a process but not available to it.

- It might have extraneous data flows, information that is of no value to any of the processes and thus needlessly complicates the interfaces.

- It might have missing processes.

- It might be incorrectly leveled.

- It might be deceptively labeled.

- It might contain inadvertent control flow or flow of control information.

- It might be conceptually incorrect. This is going to be the toughest kind of error to uncover, the kind where a DFD is demonstrably workable and internally consistent but just doesn't reflect reality.

The subsections below document a set of more or less mechanical tests to detect and remove such errors.

9.1.1 DFD Bookkeeping

It seems trivial, but the most common of all DFD failings is the unnamed data flow. Nearly every time you "forget" to name a data flow, there is a gross error hidden in your DFD. When you try to put a name on the data flow, you realize that it is unnameable. *A data flow is unnameable for one and only one reason: because you haven't got the foggiest idea of what information it carries.* Unnamed and unnameable data flows are most often the result of concentrating first on the "functions," whatever they are, rather than on the pipelines of data. Fig. 48 is an extreme example of this problem. It looks fine until you try to specify the interfaces. I suggest that the diagram in Fig. 48 does not accomplish the requirement of a DFD; it does not *partition* in any meaningful sense.

Fig. 49 also shows some unnamed data flows. The only difference here is that some meaningless labels have been drawn alongside the vectors to conceal the fact that nobody knows what information is supposed to flow along them. I suggest that terms like data, information, input, and output have so little significance on a Data Flow Diagram that they ought to be banished.

The ultimate test of whether the composition of a data flow is known is the analyst's ability to define it by writing a Data Dictionary entry. In order to do this, he has to know the sum total of all information that flows over the pipeline and all the possible variants of it.

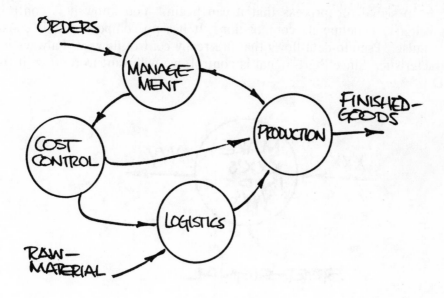

Figure 48

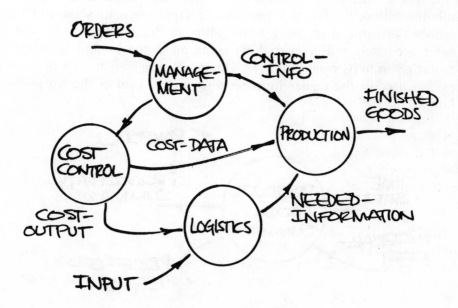

Figure 49

This definition process will also help you eliminate pseudo-data flows like the one called Start-Signal in Fig. 50. Start-Signal is a control item. Its only purpose is to signal the process that it can begin. You know it is control as soon as you try to define its composition. It has no composition. It consists only of a pulse. Pseudo-data flows that are really control flows will always have this characteristic. Since Start-Signal is control, we will want to remove it from the DFD.

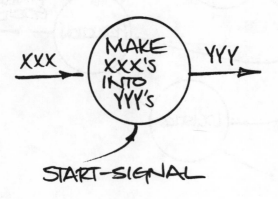

Figure 50

There are two good tests for bubble names that you ought to apply. First, does the name make sense in terms of the input and output data flows associated with the bubble? (Try that test on the DFD segment shown in Fig. 51. What does the name there have to do with the Price-of-Eggs?) Second, does the name accurately reflect everything going on in the bubble as indicated by the names given to its child bubbles? Obviously, what we are trying to avoid is names that deceive the reader by concealing much or all of the work allocated to a bubble.

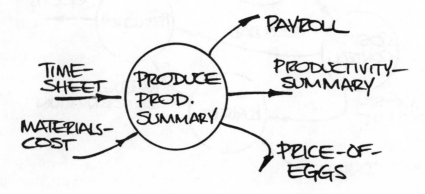

Figure 51

In summary, your rigorously applied DFD bookkeeping standard ought to include the following:

- Make sure every data flow has a name.

- Make sure you can define every name.

- Eliminate "data flows" with null composition.

- Test bubble names against inputs and outputs.

- Test bubble names against the lower levels.

9.1.2 Consistency Errors

A common failing is that DFD's are self-contradictory or have holes or redundant processes. Usually, you will run across such problems as you work your way down to the bottom levels. Make sure that whenever you add or delete data flows at lower levels, you keep your DFD set in balance. In particular, when you add a new data flow at level n, go back up to the parent and figure out where it has to come from or go to. If you forget to do this, you may miss a change that has to be made in another area. Careful application of the balancing rule is your best means of keeping the results consistent.

In analyzing the current system, be particularly careful of your description of work that falls on or near the boundary between two different users' areas of responsibility. Frequently both users consider the boundary process to be part of their areas, so you end up with the same work shown in two places. This will almost always show up as a balancing problem as well. Resolve it by putting the single process in either one place or the other. The opposite situation can also occur: a boundary process that neither user remembers to tell you about. Usually that will be signalled by an output data flow from one area not being exactly equivalent to the input you expected in the other.

9.1.3 Data Conservation

Look at the DFD segment shown in Fig. 52. Something is obviously wrong, since some of the information necessary to prepare a Complete-Invoice is not shown flowing into the process. There is no customer name, for instance. If you know (from the Data Dictionary) that the Complete-Invoice might have to show an amount of prepayment, then that, too, is missing. In order to do its work, the bubble has to be a net source of customer names and prepaid amounts, something it could not possibly be. I refer to this as a *data conservation error*. Information is required to flow out of a process that did not flow in to it in any form.

The data conservation rule that we ought to apply is this: A process must be able to build its outputs using only the information in data flows explicitly shown flowing into it, plus constant information. Processes that violate this rule are incorrect. They imply that at least one interface has been forgotten. Whenever this occurs, you have to go back and add in the required data flow in order to make your diagram correct.

You may also encounter the opposite situation in which a data flow or some component of it dies inside a process. This is not a guaranteed error, but you have to be suspicious of it. After all, why does your user currently route that information to the process if it is of no use? The interface would be simpler without it. Since the user, in his management of manual procedures, is as concerned with simple interfaces as we are, chances are he would have removed the offending item if it were really not required. It is a good rule to go back to the user whenever a dead-end data flow is detected and query him about it.

Later in analysis, when we are describing the new system rather than the current environment, dead-end data flows will be considered true errors. We will then want to remove all useless data flows in order to take advantage of simplified interfaces.

9.1.4 File Problems

Fig. 53 shows two diagrams of a leveled set. There is an obvious error in the figure. Can you spot it?

The error is that the file called Delta is an information sink: Data goes into it but never comes out. You know this because of a by-product of the balancing convention, the idea that on the first level where a file is declared, all references to it are shown. Delta is local to Diagram 4, so all references to it are shown there. Diagram 4 says that nobody ever uses the file. Whenever you see this kind of an unused file on one of your DFD's, consider it a sign of missed communication with your user. After all, he is no jerk; he doesn't keep the file around for no purpose whatsoever. When you ask him what he uses the file for, you will almost certainly uncover some function that you had not previously suspected.

Remember that the data flow connecting a bubble to a file must show the *net* direction of flow. For instance, you would not show a double-headed arrow between Bubble 4.1 and Delta even though the bubble has to read the file in order to update it. (The double arrow would only be used if the process updated the file *and* used information from it for some other purpose.) I stress this point here because if you don't follow this convention, errors such as the one in Fig. 53 will not be evident.

9.1.5 Conceptual Errors — The Analysis Phase Walkthrough

The only way to get rid of conceptual errors is to work closely with the user and his staff, testing out the DFD's for accuracy in their representation of business operations. Your major tool in such an effort is the walkthrough. Chances are you are already using walkthroughs to some extent in your organization. If not, you might profit from a reading of Ed Yourdon's *Structured Walkthroughs*, cited in the Bibliography. It makes the case for the use of walk-throughs and makes some good suggestions about how to run them.

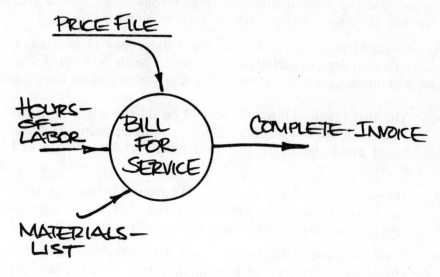

Figure 52

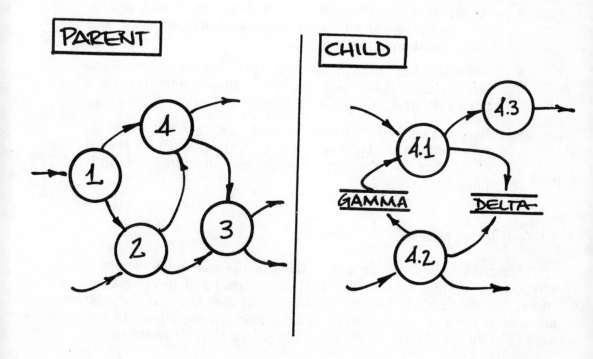

Figure 53

Walkthroughs used during the analysis phase are rather different from those in the implementation phases. The questions considered in the analysis phase walkthrough are of more critical importance to the project's success. The user is present, so the political climate is more complicated. The subject matter lies on the boundary between two areas of responsibility: yours and his.

As you might suspect, conduct of the walkthrough is affected by these differences. To complement techniques you may already have settled upon for managing walkthroughs, I offer this list of special analysis phase considerations:

- The first few walkthroughs should be on the users' own turf. They ought to last no more than ten minutes each. The users ought not to be aware that they have participated in a walkthrough.

- Do not send out DFD's to users in advance unless they are already familiar with the methods in use. The users' introduction to new methodology ought to be under your strict control.

- Show new users the middle- and lower-level diagrams first. Hold back the concept of leveling until they have become familiar with the basic ideas.

- Complement your walkthrough of the DFD with additional physical information (people, places, document names, and so forth) to help the user follow along.

- As the Data Flow Diagrams become more presentable, and the users more familiar with them, hold formal walkthroughs. Use an overhead projector, if possible.

- While you walk through a level n DFD, members of the walkthrough team should have level n-1 in their hands. This helps them keep track of the context of your discussion.

- In all dealings with the user, avoid the use of jargon specific to your methods. A DFD is a picture. A Data Dictionary is a list of interface and document descriptions. Structured English is English.

9.2 Tests for usefulness

Sometimes you come up with a DFD that is technically correct but not very useful. By this I mean that it is hard to read and to comprehend. Of course, the ultimate test for this problem is the very fact that the diagram is unreadable, but it is often difficult to apply this test objectively to your own work. I make use of three more or less mechanical tests for readability:

- interface complexity
- process names
- evenness of partitioning

9.2.1 Interface Complexity

The number and clarity of data flows into and out of a given bubble has a strong effect on the usefulness of your partitioning. I won't try to set any standard for what the maximum number of data flows in and out ought to be. But this rule always seems to apply: *Less is better.*

Whenever you go back to a set of DFD's for a subsequent iteration, look first at the most complicated bubble — the one with the most interfaces, the most compound interfaces, and the greatest variation in type of interfaces. Start your improvement process right there. How could that bubble be changed to cut down on interface complexity? What portion of its work could be spun off and allocated to another bubble in such a way as to reduce passed information? Perhaps the whole process should be split into two or three pieces, or amalgamated into some other bubble. Anything you can do to decrease complexity of interfaces at a given level is an improvement.

9.2.2 Process Names

After you have given each bubble the most accurate descriptive name you can devise, go back and listen to what the names are telling you. All names are not equal. Consider these process names, for instance:

1. Calculate Commission Amount

2. Produce Morning Stock Summary

3. Edit Transaction and Debit Customer Record

4. Process Order

5. Handle Input

6. Do Miscellaneous Stuff

The first two of these are strong names. They consist of strong action verbs, each one coupled with a single explicit object. Chances are, they are functional primitives with a single input and a single output. The third name is not too bad, but you might consider breaking the bubble in two if you have room.

Names 4, 5, and 6 are all more or less awful. The word "process" must certainly be one of the wishy-washiest words in the English language. Its meaning is essentially this: Look at the object of the sentence, and do whatever sorts of things you usually do to objects of that type. It dodges the function of the verb entirely. "Handle" is just as bad. The name Handle Input is a double atrocity. Not only is the verb a cop-out, so is the object. Input doesn't mean anything. It is a general-purpose non-specifier used to refer to odds and ends heading toward the center of a system. Handle Input is little better than Do Miscellaneous Stuff. You might as well label your processes with question marks.

Weak process names are signs of poor partitioning. When you detect them, try to repartition to come up with more nameable sets of work. Don't waste your time trying to think up a good name for a bubble that really can only be called Handle Input or some such thing — get rid of it.

The ideal name is one that consists of a single strong verb with a single concrete object. But you will only be able to come up with such names near or at the bottom of your hierarchy (where you should insist on it). In fact, a very strong name is a good sign that you have reached the bottom, that the bubble you are dealing with is a primitive. At the upper levels, you will be hard-pressed to think of such single-minded labels, but there is a world of difference between the names you can come up with for usefully partitioned sets and the best ones you can honestly give to sets of work that don't belong together. If you feel uncomfortable with the name of a given bubble, try to break it apart, distribute all or part of its work to other bubbles, or include it someplace else. Each time you do this, go back and rename the affected bubbles and then see whether you have improved or worsened the diagram.

9.2.3 Uneven Partitioning

The ideal partitioning is one that divides into *even-sized* pieces. Of course you cannot hope to do that exactly, but you can avoid the extremes of uneven-ness. If one of your bubbles on a given level is primitive, and another needs to be partitioned down three or four more levels, you have not achieved a useful partitioning. You have just "chipped" off an insignificant piece and left the rest. Readability will suffer because the diagram will combine some details and some higher levels of abstraction. Go back and try again.

9.3 Starting over

A mechanical approach to starting over is something called "topological repartitioning." Rather than trying to describe this idea with words, I will rely on pictures. Fig. 54 shows a partial set of leveled DFD's that are badly in need of topological repartitioning. Fig. 55 shows the repartitioned set.

Whenever a lower-level diagram turns out to consist of disconnected or substantially disconnected networks, you have a candidate for this kind of repartitioning: the parent. You might go about it like this:

1.   Build an expanded diagram by combining all of the children of the diagram that needs repartitioning. Connect the interfaces. Unless the set was out of balance, you will always be able to do this.

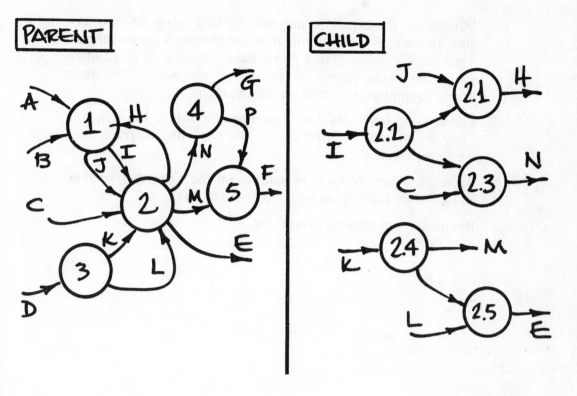

Figure 54

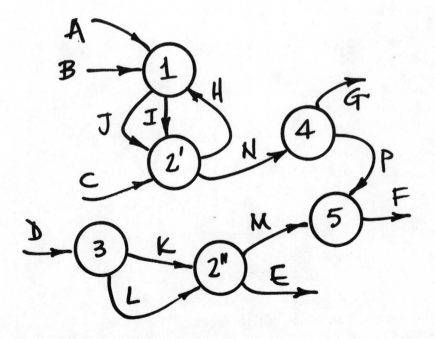

Figure 55

2.   Working on the expanded diagram, try to divide it into a workable number of subsets with minimal interfaces among them. Your decision to include a given bubble in one set or another should be made purely on topological grounds — in other words, to minimize interset communication.

3.   Recreate the upper level by allocating one bubble for each set. Copy the interset interfaces from the expanded diagram onto the upper level.

4.   Recreate the lower level by cutting up the expanded version along the set boundaries with a pair of scissors.

5.   Renumber and rename everything.

10 DATA FLOW DIAGRAMS FOR SYSTEM SPECIFICATION

You cannot consider the analysis phase over until the major interface between people and machines has been established and documented. The Structured Specification that describes our system ought to make this interface evident to the reader and help him work out potential problems with it.

There are two possible ways to incorporate information about the man-machine interface into the Structured Specification: We might document the dialogue directly, using a DFD, or somehow group our DFD's so that the interface stands out.

10.1 The man-machine dialogue

Fig. 56 shows the first alternative, use of a Data Flow Diagram to document a man-machine dialogue. Since the dialogue really does consist of data flowing between manual and automated procedures, the DFD is an ideal vehicle to present it.

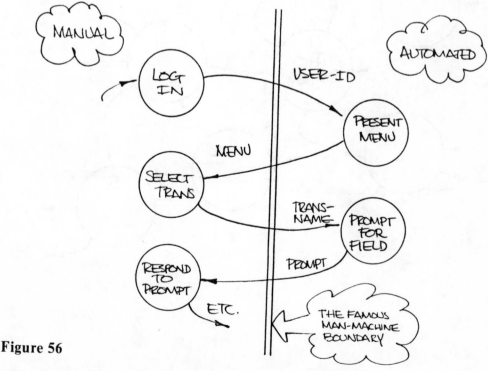

Figure 56

Now the Structured Specification will consist of a set of leveled DFD's and supporting documentation plus an extra set of DFD's to portray the man-machine dialogue. The trouble with such an approach is that it necessarily introduces redundancy — the border processes are described in two places. Any change to the boundary itself or to processes at or near it will require multiple updates to the Structured Specification. You may be willing to live with this since the DFD's explicit statement of the dialogue is a great asset during analysis. But there is another possibility, the integrated top-level approach.

10.2 The integrated top-level approach

Fig. 57 shows the man-machine interface documented in a slightly different fashion. Here we achieve our end not by creating a new interface document, but by regrouping the DFD's of the Structured Specification in order to declare the interface between man and machine. Specifically, all man-machine interaction has been forced up into the top level.

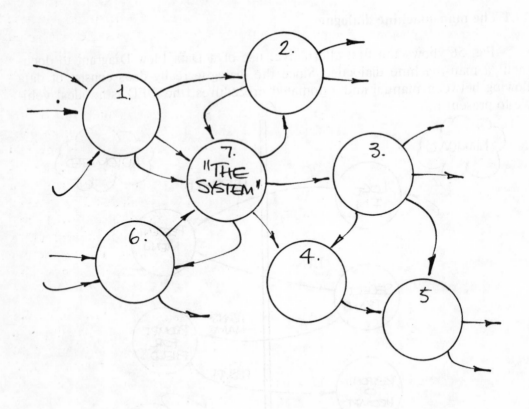

Figure 57

10.2.1 The Man-Machine Boundary

The automated system is shown in Fig. 57 as a single bubble, Bubble 7. So the man-machine boundary is the periphery of that bubble. The interface between human and automated procedure is the set of data flows moving into and out of Bubble 7. According to the balancing convention, there is no additional interface between the automated system and any set of manual procedures introduced at lower levels.

It follows that Bubble 7 and all of its child and grandchild figures, all the way down to the bottom, plus the associated Data Dictionary and mini-spec describe *automated* procedure. Bubbles 1 through 6, plus all their lower levels and support documents, describe *manual* procedure. The Structured Specification is made up of the full set, both manual and automated. The same tools are used to describe both.

10.2.2 Documentation of Manual Procedures

You may be surprised to consider such early and elaborate documentation of manual procedures as is suggested by Fig. 57. However, Structured Analysis requires that the design of manual patterns for the system's use be completed and documented *during the analysis phase*. This is a substantial departure from the classical approach, where procedures for the use of the system are not drawn up until the very end of the project when user manuals are written.

The reason for moving manual procedure design up to the front of the life cycle is the following: The user is at a disadvantage in any discussion that deals exclusively with automated process. The only thing that makes any real sense to him is the way that he and his people are going to make use of the automated system. All too often, when you ask a user for his concurrence on a classical specification (consisting of a description of the insides and boundaries of some computer system), he reads through it and tries to relate it to an implied set of manual procedures. Then he applies that set of manual procedures to his own environment and decides whether or not it is workable. To the extent that the procedure implied to him in his reading of the specification was invalid, his sign-off on the document is useless.

Since users feel more comfortable evaluating manual procedures for working with the system rather than evaluating the system itself, we have to make those procedures an *explicit* part of the Structured Specification. We do that by extending the context of our study well beyond the man-machine boundary, out far enough so that the Structured Specification includes a description of all manual procedures related to the installation of the system.

If your management gripes because this draws out the analysis phase, use this set of justifications:

- The work has to be done anyway. Doing it earlier will not adversely affect final delivery.

- Documenting human procedure during analysis helps to elicit *meaningful* concurrence from the user.

- Insofar as the user is more aware of how the system will affect his environment than he might otherwise have been, acceptance and parallel testing will be smoother. There will be fewer surprises for everyone.

10.3 Problems and potential problems

I'm sure there are questions that are bothering you. Below are some of the questions most commonly brought up in my seminars, together with my answers. I hope that I have managed to hit upon the ones that you might ask.

How is a Data Flow Diagram different from what we used to call a system flowchart? DFD's differ from system flowcharts in two important respects. First, DFD's document the flow of data, not the flow of control. The system flowchart is still a flowchart — it records the stream of consciousness of the data processor. Second, the system flowchart could never be used as a specification tool because it tells flagrant lies. The flowchart's presentation of the overall philosophy of a system ignores anything that is not a major part of that philosophy; i.e., it pretends that such considerations simply do not exist. The DFD, on the other hand, is rigorous — it presents the philosophy at the top, and the details at the bottom, with a smooth progression from abstract to concrete.

Don't the top-level diagrams get terribly complicated? Sometimes the top level is cluttered, and it often has considerably more than seven bubbles. Sometimes (horrors!) it even has some crossed data flows. My experience has been, however, that you can always draw a workable top level, one that both analyst and user can understand and find useful. Remember that the top-level diagram is the one you use most. So you build familiarity with it. The kind of understanding-at-a-glance characteristic that you learn to expect at lower levels does not usually apply to the top, but it is not as necessary there.

Don't users balk at working with DFD's? Many users are already familiar with some form of network descriptive techniques, so DFD's don't seem so foreign to them. Users who have never worked with paper flowcharts or PetriNets are more of a problem. They require some special handling. How you present the idea to them is most important. Try this: While you are interviewing a user, sketch out a DFD of the area he is describing. Show it to him, and ask if you've got it right. Don't give the diagram an overbearing name — if he asks you what it is, tell him it is a picture of his operation.

Don't introduce the idea of leveling until he has seen a number of unleveled DFD's. His first contact with leveling ought to be downward; for some reason, leveling downward is conceptually easier to grasp. Only when the user is completely conversant with the techniques, should you show him the top level. Introduce this by telling him you have prepared an overview that shows how his whole area fits into the context of the company's operations.

Isn't it often difficult or impossible to get started on a pure top-down analysis? Yes. The problem is that, in a large enough operation, nobody sees the big picture. In such a case, you might have to start in the middle, where there is some expertise. When you have collected all the middle-level pictures, combine them into one enormous diagram. This is your top level. Of course, it is unreadable. Now you have to construct a level to go above it. (The process of doing this is very like what I called topological repartitioning in the previous chapter.) The resultant document is pure top-down.

How many levels should we expect? It depends on the size of your system, and the extent of your partitioning at each level. I have never seen leveled DFD's go beyond ten levels, for the simple reason that ten levels allows for so many primitives that the largest system on earth wouldn't need more.

When you look at the details of level n, don't you find that you have to go back and modify level n-1? Yes, it happens all the time.

Doesn't that sometimes invalidate the partitioning at the upper level? Yes. You add one more input or output data flow into the bubble on the upper level — because you discover from your study of the next level down that it is required — and it is the straw that breaks the camel's back. Burdened with yet another interface, the higher-level partitioning is just unacceptable. So, you back up a level and start over.

Isn't it possible that you might find something at the very bottom, or near the bottom, that would invalidate everything all the way to the top? I suppose it is technically possible, but I have never seen it happen. Most often the ripple effect only goes up one level.

You are very sanguine about ripping up a DFD and starting over. Don't you waste an incredible amount of time doing this? Honestly, it is not very much time. It is, after all, only the artwork that gets thrown away, not the ideas. And I don't consider the time wasted either, since we make significant progress with each iteration.

Is it OK to have several different names for the same data flow? Yes, the extra names are called aliases. It is often helpful to use aliases (when two different users refer to the same document or file by different names, for instance), but you musn't allow them to proliferate. Be particularly severe about aliases that are introduced by the analyst inadvertently. If you call a data flow Requisition-for-Parts in one place and a Parts-Requisition someplace else, everyone will know what you mean; but you still have to make two entries in the Data Dictionary. If you do this too many times, the Data Dictionary will become unworkable, and you will have to waste valuable time weeding out the aliases.

How do you stick a set of leveled DFD's together? What order do you keep them in? There is no completely satisfactory answer to this question, since any ordering method is inherently one-dimensional, and we need two-dimensional access. Making a separate loose-leaf binder for each level, and keeping diagrams in numerical order in the binder allows you to work with several levels at once without having to page back and forth.

Do you ever portray flow of physical goods on a DFD, or are you restricted to pure data? It is often difficult to separate the two. A pile of cast parts, for example, is clearly not data. It is stuff. But it has data content. A worker may be able to look at the pile and tell you the part numbers of the items there. He can count the pieces to tell how many there are. In many cases, the data content of physical goods is essential to an operation. When this is true, you definitely need to show physical goods on the DFD. There may even be people shown on a DFD, although this is much rarer. As an example, if you were doing a study of a hospital, you might be obliged to show the patients as data flows; hospitals are continually extracting data from patients — their ages, blood types, and pulse rates. The doctors and nurses would not show up on such a DFD, of course, since they are processors, not data.

Are you serious about never allowing crossed data flow lines on a DFD? Not terribly. Sometimes they are unavoidable. If you get too many, however, nobody will be able to understand your diagrams. A little common sense in redrafting will get rid of the worst offenders. Too many bubbles on a given level will make the problem more serious. (You can avoid crossed lines altogether by limiting your partitioning to two bubbles at each level, but this seems extreme.)

PART 3

DATA DICTIONARY

11 THE ANALYSIS PHASE DATA DICTIONARY

As you might suspect, much of what follows is concerned with definitions. For starters, we need a definition of Data Dictionary (DD); here is one from James Martin. [1]

A Data Dictionary is a repository of data about data.

In addition, Data Dictionary is usually understood to include the set of procedures used to build and maintain the repository.

A set of a dozen or so definitions would never be called a dictionary; it would be called a glossary. The fact that I am using the term dictionary here implies that the set of definitions to be maintained during a Structured Analysis project is large — and it is. Think back to the case study example in Chapter 8. It dealt with only a small part of a reasonably small operation. Yet even a partial Data Dictionary to support the set of Data Flow Diagrams presented there involved a few hundred entries. It is not at all unreasonable for a good-sized project to involve several thousand data flows. Together with all the subordinate data flows, files, data bases, and odds and ends that would have to go into our Data Dictionary during the analysis phase of that project, we could easily end up with 5,000 entries. We need a very well-organized approach to cope with such a monster. That well-organized approach is the subject considered here.

In addition to Martin's definition presented above, I think it would be useful to present an "undefinition," a statement about what Data Dictionary is not:

Data Dictionary, as we use the term in Structured Analysis, does not necessarily refer to any of the commercially available packages that help you keep track of physical data descriptions during implementation.

Such packages generally date from the mid-1960's, long before the advent of Structured Analysis. I will have more to say about them in Chapter 14. But Data Dictionary is a much larger topic than just the packages.

---

[1] James Martin, *Principles of Data-Base Management*, Prentice-Hall, 1976.

11.1 The uses of Data Dictionary

The DD is an integral part of the Structured Specification; without it, Data Flow Diagrams are just pretty pictures that give some idea of what is going on in a system. It is only when each and every element of the DFD has been rigorously defined that the whole can constitute a "specification." The set of rigorous definitions of all DFD elements is the Data Dictionary.

The most important role of any dictionary is to give you a single place to look up definitions of terms you do not understand. That is precisely the role that the Data Dictionary is going to take on during Structured Analysis. It is mainly composed of definitions, definitions of

- data flows
- components of data flows
- files
- processes

and anything else that needs defining. Many DD users freely include definitions that might otherwise go into a project glossary. They reason that since there is already a set of procedures to keep track of and maintain definitions, why not use it for everything?

On the same theory, analysts often include extraneous information about each item right along with its definition. For instance, as they learn more and more about a given data flow, they include information about

- frequency
- volume
- size
- affected users
- peaks and valleys
- security considerations
- priority
- implementation schedule

and so forth. All of these are in addition to the primary entry, the definition.

Many of the items cited above are not even known during the analysis phase. Mentioning them implies that the Data Dictionary ought to be expected to survive the analysis phase, and to go on and serve some function in later project phases. Although I do intend to imply this, it should not be overemphasized. The principal use of the DD is as an analysis phase tool, and its use in the analysis phase is almost exclusively for keeping track of definitions.

11.2 Correlating Data Dictionary to the DFD's

Data Flow Diagrams and the Data Dictionary have to be considered together. Without a DD, the diagrams lack rigor; without the diagrams, the DD is of no use to anyone. The correlation between the two is as follows:

There is one Data Dictionary entry for each unique data flow that appears anywhere in the DFD set. There is one DD entry for each file referenced on any diagram in the set. There is one DD entry for each functional primitive in the set.

There may be other entries as well. In particular, it may be useful to introduce and define some subordinate data flows — data flows which do not actually appear on any DFD, but which are handy to use as components in the definitions of data flows which do.

11.3 Implementation considerations

How shall we organize our Data Dictionary? I know of three possibilities, all in common use today:

- *Totally manual procedures,* making use of index card files, loose-leaf notebooks, and the like, plus a considerable amount of clerical support.

- *Totally automated procedures,* involving one of the commercially available DD packages (possibly modified to make it more tailored to the special analysis phase requirements).

- *Combined manual and homegrown automated procedures,* taking advantage of whatever support facilities are available in the organization (text editors, report generators, and so forth).

Regardless of the approach taken in implementing DD procedures, the following requirements apply:

- Definitions must be readily accessible by name.

- There should be no redundancy in the DD.

- There should be little or no information in the DD that is already contained in some other component of the Structured Specification.

- It must be simple to make updates to the DD.

- The convention for writing definitions should be straightforward and orthogonal.[2]

In addition, there are some "bells and whistles" that might come in handy. I include in this category facilities to generate cross-reference listings, error detection, consistency checking, and the like.

───────────

[2]A convention is orthogonal to the extent that it provides one and only one way to define a given term. Check the Glossary for a more complete definition.

12 DEFINITIONS IN THE DATA DICTIONARY

A dictionary is an ordered set of definitions; a Data Dictionary is an ordered set of definitions of terms used in a DFD. But what is a definition? The kinds of "definitions" we are accustomed to using in implementation are clearly not relevant here. It is much too early in the project to make statements such as:

> RAZAFRAZ is a six-character alphanumeric field, right-filled with blanks.

And even if we were ready to establish such physical attributes of the various data names, they just don't constitute definitions. A definition is something more . . .

12.1 Characteristics of a definition

There is no way to begin this section except with a definition of the word definition:

> A definition is a description consisting of *genus* and *differentia*. The genus establishes some class that contains the word being defined, and the set of differentia distinguishes it from all other members of that class.
>
> — Adapted from Aristotle

Aristotle's own example is his definition of Man: "Man is the animal possessed of the capacity for articulate speech." Genus — animal. Differentia — "possessed of the capacity for articulate speech."

I hope you won't think me pedantic for quoting Aristotle in this context. His approach to defining terms is particularly appropriate to Structured Analysis, since it neatly partitions the problem into two pieces. All we have to do to define a data flow, for instance, is determine some class it belongs to, and then establish a set of distinguishing characteristics in order to leave no doubt about which member of the class we mean.

12.1.1 Classes of Items to be Defined

We can make do with a very limited number of classes in our Data Dictionary. They are the obvious ones, the classes that describe the components of Data Flow Diagrams:

- data flow
- file
- process
- data element

You have already encountered the first three of these.[1] The term "data element" is often used without a precise definition on the theory that it means just what the name implies, an element of data. Unfortunately, that doesn't say much. (An "element" is more or less a "thing.") As I use the term

A data element is a special kind of a data flow, one that cannot be decomposed into subordinate data flows.

As you will see, defining data flows will frequently involve declaring them to be composed of other data flows, lower level ones. Data flows which cannot be represented in this fashion, as amalgams of other data flows, are data elements. Data elements are the "primitives" of our definition process.

All the relevant terms are thus defined, with the possible exception of data. "Data," as we all know, is the plural of "datum," and a datum is . . . a little wordsy-numbersy thing. I am not being entirely facetious in dodging the task of defining a datum. Any definition process must stop somewhere. For the purposes of this book, "datum" is a self-defining term. More about self-defining terms in just a bit.

Having established the four classes of things to be defined, all that is left in our definition process is the differentiation part. The means of differentiation will depend on the class of item being defined.

12.1.2 Top-Down Partitioning of Data

Most definitions accomplish their purposes by representing the item being defined as some combination of components (e.g., Definition = Genus + Differentia). The components themselves are then defined as being composed of lower-level components, and so forth. In this sense, a definition is a top-down partitioning.

---

[1]To refresh your memory: A *data flow* is a pipeline over which data of known composition is transmitted. A *file* is a time-delayed repository of data. A *process* is a transformation of incoming data flow(s) into outgoing data flow(s).

Definitions in the Data Dictionary are top-down partitionings of data. If we know that data flow A is composed of one B and one C, and that B is made up of B1 and B2 and B3, while a C is always C1 and C2, we could write the definition in a single step as

$$A = B1 + B2 + B3 + C1 + C2$$

but we probably would not. Particularly when a data flow is complex, it makes sense to define it in terms of meaningful high-level subordinates, and then define those subordinates. To get a complete understanding of what the term means, a reader might have to look up several entries in the Data Dictionary. But the definition would move smoothly from the most abstract to the most detailed. It would not bury the reader immediately with the lowest-level components, all 62,341 of them. If the upper- and mid-level subordinate data flows used in definitions are aptly chosen, readers will quickly build a working vocabulary of them, and thus be able to use the Data Dictionary with great efficiency.

What are these "components" that are used to define entries in our Data Dictionary? Usually other entries in the Data Dictionary. A data flow, as an example, might be defined in terms of other data flows and data elements. A file might be defined in terms of data flows and data elements. A file might also be composed of subordinate files. Such a file would be more commonly called a data base. (It is still a "file" for our purposes, a time-delayed repository of data.) A process may be made up of subordinate processes.

12.1.3 Relational Operators

Most Data Dictionary definitions are formulas that declare the thing being defined to be made up of some *related* set of components. An example is the following:

Task-Designator IS EQUIVALENT TO: EITHER: Job-Step-Number

OR: Owner-Code AND
Task-Name

The kinds of relationships expressed by the all capital letter words in the preceding example are relational operators.

Our requirement for DD is to select some set of relational operators that will allow us to define any possible dictionary entry. In going about this, it is desirable to choose a small set of simple operators, and then allow them to be used in combination to construct complex operators. This will spare us the necessity of cluttering up our definitions with such horrors as "EITHER OF OR NEITHER OF BUT NEVER BOTH OF."

Bohm and Jacopini (see Bibliography) presented an elegant proof that any process or program can be made up elements related in only three ways:

1.   a sequential set of instructions, executed one after the other

2.   a closed-end decision construct (CASE or IF-THEN-ELSE)

3.   a closed-end looping construct (DO-WHILE or REPEAT-UNTIL)

It turns out that these same three relationships will serve to describe data as well as process. We need only to rephrase them a bit to formulate our basic operators:

1.   *Sequence,* the concatenation of two or more components in order. (The AND in the Task-Designator example above serves as a sequence operator.)

2.   *Selection,* the choice of precisely one of two or more possibilities. (EITHER and OR together served as a select operator in our example.)

3.   *Iteration,* the repetition of a designated component zero or more times.

An example of the iteration operator is:

Flight-Manifest IS EQUIVALENT TO: ITERATIONS OF: Passenger-Name

These three operators can be used to define any entry in our Data Dictionary. The fact that these three alone are sufficient for our needs is not immediately obvious, but it can be demonstrated with exactly the same logic that makes up the Bohm and Jacopini proof. Rather than reproduce that for you here, I will try to give you enough examples to dispel any doubts.

The iteration operator is often used with upper and lower limits to specify how many times something can be iterated:

Label IS EQUIVALENT TO: 1 TO 8 ITERATIONS OF: Character

By establishing the lower limit as zero and the upper limit as one, you can express the idea that a component is optional. I find this a rather foreign way to describe a very familiar kind of relationship, so I propose the following extra operator:

4.   *Optional,* meaning zero or one iteration(s) of a component.

An example would be:

Order IS EQUIVALENT TO: Mailing-Coupon AND
OPTIONAL: Prepayment

In summary, a compact set of relational operators that can be used to construct any definition in our DD is the following:

- IS EQUIVALENT TO
- AND
- EITHER-OR
- ITERATIONS OF
- OPTIONAL

These operators will not only be technically sufficient for our requirement — but, by adroit use of them, we should also be able to come up with definitions that are meaningful and conceptually easy to understand.

12.2 Definition conventions

The practice of spelling out the operators in the form suggested in the previous section is obviously too cumbersome for our needs. Instead, I propose the following notation:[2]

$=$ means IS EQUIVALENT TO.

$+$ means AND.

[]   means EITHER-OR; i.e., select one of the options enclosed in the brackets.

{}   means ITERATIONS OF the component enclosed.

()   means that the enclosed component is OPTIONAL.

The iterations braces are often annotated with upper and/or lower limits. Some analysts write the limits as superscripts (for upper limit) and subscripts before the left-hand brace. Others put the lower limit to the left of the opening brace and the upper limit to the right of the closing brace:

$$\substack{5\\1}\{x\} \text{ is the same as } 1\{x\}5$$

I shall use both conventions in this text.

Armed with this abbreviated notation, let's go back and rewrite the examples presented in the previous section:

$$\text{Task-Designator} \quad = \quad \begin{bmatrix} \text{Job-Step-Number} \\ \text{Owner-Code} + \text{Task-Name} \end{bmatrix}$$

---

[2] Adapted from Bacus-Nauer Form.

$$\text{Flight-Manifest} \quad = \quad \{\text{Passenger-Name}\}$$

$$\text{Label} \quad\quad = \quad {}^{8}_{1}\{\text{Character}\}$$

$$\text{Order} \quad\quad = \quad \text{Mailing-Coupon} + (\text{Prepayment})$$

Some analysts like to annotate their definitions with comments. A convention for this is to enclose the comments in asterisks:

$$\text{Paid-Amount} \quad = \quad \begin{bmatrix} \text{Dollar-Amount} \\ \text{Pound-Amount *UK residents only*} \end{bmatrix}$$

12.2.1 Use and Misuse of the Conventions

The select (EITHER-OR) operator seems to give people the most trouble. Note that the square brackets are always used to include two or more options. Usually the options are piled one on top of the other as follows:

$$\text{Whatever} \quad = \quad \begin{bmatrix} \text{Option-1} \\ \text{Option-2} \\ \text{Option-3} \end{bmatrix}$$

Alternately, the options might be separated by a |, as in the following:

$$\text{Whatever} \quad = \quad [\text{Option-1} \mid \text{Option-2} \mid \text{Option-3}]$$

Square brackets enclosing a single component have no meaning.

The iteration operator says that everything enclosed within the braces is repeated. For instance, you might have a portion of an invoice expressed in this way:

$$\text{Invoice-Body} \quad = \quad \{\text{Invoice-Line}\}$$

$$\text{Invoice-Line} \quad = \quad \text{Quantity} + \text{Item-Number} + \text{Unit-Price} + \text{Item-Subtotal}$$

or you might choose to define the same thing in this fashion:

$$\text{Invoice-Body} \quad = \quad \{\text{Quantity} + \text{Item-Number} + \text{Unit-Price} + \text{Item-Subtotal}\}$$

The two definitions are equivalent. Sometimes new users of Data Dictionary try to invent an AND-OR operator by piling up possibilities inside the iteration symbols — that has no meaning in the convention used here. If you need to express such a thing as "one or the other or both," just listen to what you are saying (that there are three options) and express it accordingly:

$$\text{A-Or-B-Or-Both} \quad = \quad [\text{A} \mid \text{B} \mid \text{A+B}]$$

The fact that the iteration operator specifically allows for zero iterations of the component tends to be a source of some confusion. Unadorned braces mean "from zero to infinity repetitions of the component enclosed." So it follows that:

$$(\{ \text{Whatever} \}) \qquad = \qquad \{ \text{Whatever} \}$$

If you want to express the idea of "one or more repetitions of something," you have to apply the lower limit:

$$_1 \{ \text{Something} \}$$

I believe that the use of combinations of operators is the most difficult aspect of defining. If you find that you have a definition that is strongly *nested*, and it is giving you trouble or seems difficult to read, consider introducing some intermediate-level components. Here is an example of a definition that I consider over-nested:

W2-Composite-Summary = {Subsidiary-Name + Address-1 +
Address-2 + Zipcode + Corporate-Taxpayer-ID +
{[Employee-Name + Social-Security-Number |
Contractor-Name + Corporate-Taxpayer-ID] +
Amount-Withheld} + Subsidiary-Total-Withheld} +
Overall-Total-Withheld

The same definition in a much more readable form is

W2-Composite-Summary = {Subsidiary-Summary} +
Overall-Total-Withheld

Subsidiary-Summary   = Subsidiary-Identifier +
{Withholding-Item} +
Subsidiary-Total-Withheld

Withholding-Item   = [Personal-Identifier / Contractor-Identifier] + Amount-Withheld

and so forth.

12.2.2 An Example

Presented below is a composite set of Data Dictionary definitions using the defining conventions. I have tried to use a "data flow" that is conceptually well understood: a telephone number. The following defines a legal dialing sequence from any Yourdon phone:

Telephone-Number   =   *Legal dialing sequence from
Yourdon inc. phones*

= | Local-Extension
9 + Outside-Number
8 + WATS-Number
0   *In-house operator* |

Local-Extension   =   *Sixth Avenue building only*

=   First-Digit +   3 {Any-Digit} 3

Where the upper and lower iteration limits are the same, there are precisely that number of iterations. In this case, there are precisely three iterations of Any-Digit following the First-Digit:

First-Digit   =   [1| 2| 3| 4| 5| 6| 7]

Any-Digit   =   [1| 2| 3| 4| 5| 6| 7| 8| 9| 0]

As I have defined it, any four-digit sequence that begins with something other than 0, 8, or 9 is a legal Local-Extension. Not all of them are connected to people, but they are still legal from the phone's point of view. The "unallocated" numbers, as far as the telephone system is concerned, are legitimate dialing sequences that mean you want to be connected to the intercept trunk ("You have reached a non-working number at Yourdon inc. . . . ").

I might have approached the definition of Local-Extension differently, by defining it to be made up of only the allocated numbers:

Local-Extension   = | 6221   *Ed*
6222   *Toni*
6225   *Tim*
6228   *Rikki*

etc. |

The thing I have called Outside-Number is really any legal Bell System dialing sequence. This will vary depending on the class of central office that serves the phone. For our area, it works as follows:

Outside-Number   =   *Legal Bell System number*

= | 0   *Operator*
Service-Code
Domestic-Number
Foreign-Number |

```
Service-Code           =  | 211  *Business office*   |
                          | 411  *Information*        |
                          | 611  *Repair*             |
                          | 911  *Emergency*          |


Domestic-Number        =  ((0) + Area-Code) +
                          Local-Number


Foreign-Number         =  011 + Country-Code +
                          Routing-Code + Local-Number +
                          (# *Touch-tone phones only*)
```

I could continue, defining Local-Number to be made up of a three-digit Exchange and a four-digit Line-Appearance-Number. But I think you get the idea. Where you stop in your defining process is purely a function of where you believe further detail would be of no value, where your user understands clearly what you mean. In the telephone number example, I could certainly have done without a definition of the component, Digit.

12.3 Redundancy in DD definitions

Since one of the goals of Structured Analysis is to build a non-redundant specification, clearly there should be no redundancy in the Data Dictionary. That means you cannot allow yourself the luxury of documenting bi-directional relationships in the DD. Consider this example:

```
Item-Line              =  *Component of Bill-of-Lading*

                       =  Quantity-Shipped +
                          Part-ID + Date
```

The definition of Item-Line declares both its subordinates and its superordinate. The information that Item-Line is a component of Bill-of-Lading is redundant since Bill-of-Lading is defined as being composed of Item-Line. The relationship is declared twice.

You may feel that the clarity introduced by mentioning the upward relationship is more than worth the small redundancy (and hence slight additional maintenance problem). But suppose that Item-Line were a component of dozens of higher-level data flows — not an unusual situation at all. In that case the added maintenance headache would be substantial. The simplest set of Data Dictionary procedures is one that allows no redundancy whatsoever.

The entire Structured Specification should be non-redundant, so there ought to be no information in the Data Dictionary that is already present in some other portion of the specification. Consider this example:

Class-Specifier   =   *Output of Process 3.8.1*

                        =   Priority + (Authorization)

The comment associated with this definition is redundant with the flow documented in the Data Flow Diagram. It should be eliminated.

Redundancy among the Data Dictionary and the mini-spec descriptions of bottom-level processes is often not so obvious:

Invoice-Line   =   Quantity-Ordered +
                          Item-ID + Unit-Price +
                          Item-Subtotal *Qty x price*

In this case, the arithmetic relationship that applies between Quantity-Ordered, Unit-Price, and Item-Subtotal ought not to be mentioned in the definition, since it must certainly be declared in the description of the process which builds Invoice-Line.

12.3.1 A Rule for Reducing Redundancy

In order to avoid redundancy in and among elements of the Structured Specification, it is essential to establish guidelines for what kind of information belongs in which portion of the specification:

- Information about the *composition* of data items (what their components are and how those components interrelate) goes into the Data Dictionary.

- Information about the *content and processing* of data items (how their values are established) goes into the process description.

- Information about the *routing* of data items goes into the Data Flow Diagram.

To the extent that you work within these guidelines, the problem of redundancy will never come up.

I'm sure you noticed that in all the redundant examples presented above, it was the comment portion that introduced redundancy. There is a reason for this: The defining conventions (relational operators) used here allow little possibility of redundancy. It is chiefly the comments that can get you in trouble. That is, unless you hit upon this hideous approach:

Output-of-Process-4.2.3 = Sum-of-Prices-Times-Quantities +
                                Ten-Percent-Sales-Tax

Names ought to reflect the meaning and importance of the item, not provide information about how it is constructed or what its source and destination are.

12.3.2 On the Value of Redundancy

It would be very handy to have much of the redundant information that I am discouraging you from putting into your Data Dictionary. You can easily imagine a situation where you would like to know the names of all the data flows which contain a given component, or all the processes that have access to that component. I have no objection to redundancy per se, only to redundant maintenance.

It may be possible to have all the advantages of redundancy without the cost. If we could build an entirely non-redundant Data Dictionary, and then submit it to some program that could sort through the definitions and tell us whatever we would like to know about the set, we would have a mechanism for *controlled redundancy.* The program could derive and tell us about all the bi-directional relationships that we elected not to document directly in the DD. It could tell us the superordinates of any data item, for instance, or all the files that contained some data element of interest. It could correlate processes to data items, or the other way around. In any event, the output from the program would be full of redundancy, much of it very useful; but the input, the part we have to maintain, would be non-redundant. Such a program would be called a Data Dictionary processor. More about that idea in Chapter 14.

Whether or not we have a Data Dictionary processor, our requirement is the following: The part of the Data Dictionary that we have to maintain should have no redundancy in it, nor should any element of it be redundant (i.e., contain information that is contained elsewhere in the Structured Specification).

12.4 Self-defining terms

Sometimes you will find that a component of a definition is *not* itself defined elsewhere in the dictionary. I shall refer to such components as "self-defining terms." For our purposes, a self-defining term is synonymous with an undefined term. Since Data Dictionaries (and all dictionaries) make extensive use of self-defining terms, and since many people find the concept troubling, let's look at it in some detail.

A self-defining term is one that can be unambiguously understood from its name alone. In the definition

> **Personnel-Description   =   Employee-Last-Name +**
> **Social-Security-Number +**
> **Age**

all the components are probably self-defining terms. You could define Age further, specifying it to be precisely two iterations of any digit, but that would be a pointless exercise during analysis. Everyone knows what Age is, so there is no value in continuing.

Note that the definition process has to stop somewhere. If you did show Age to be composed of two iterations of Any-Digit, and then defined Any-Digit to be any digit — 1, 2, 3, . . . 9, 0 — you would simply have chosen to stop somewhat later. But you would have stopped. Obviously you are not going to define "1" — you couldn't in any rigorous fashion. My suggestion is that you stop when everyone on the project understands what a term means without further definition.

To the extent that you make a proper decision on what self-defining terms to use, and you name them clearly, no one will notice that they are not defined. No one will be the least bit inclined to look them up. As proof of this, I cite the fact that any English language dictionary (any natural language dictionary) is full of self-defining terms. Did you ever notice one there? I'll bet you didn't. They were so cleverly chosen that you have almost certainly never tried to look one up.

Self-defining terms in natural language dictionaries are signalled in one of the following ways:

- A term used in some other definition is not itself defined in the dictionary. (Sand, is defined as SiO_2, but SiO_2 is not defined.)

- A term is defined in a circular fashion with the term itself included in the definition. (Look up "eleven" in your dictionary.)

- A term is defined in an indirect circular fashion. ("Exist" is defined as "to be," and "be" is defined as "to exist.")

In a Data Dictionary, self-defining terms will be of the first type only.

Fig. 58 shows a tracing of some words in an English dictionary. I include it here for two reasons: because it illustrates that any dictionary (or DD) can be thought of as a connected set; and because it demonstrates the occurrence of self-defining terms in natural language dictionaries, in case you still have doubts on that subject.

Fig. 58 was constructed by looking up the (randomly selected) target word, and forming a two-word definition from the largest genus and smallest differentium given there. I wrote the genus ("meal") on the genus axis, and the differentium on the differentium axis. Now I continue by looking up those two words, plotting the genus of the genus and the differentium of the differentium, and so on. As you can see, moving along the genus axis, you encounter more and more general classes of words (since each one includes the previous). Moving along the differentium axis, the words become more and more specific. In either direction you finally encounter a self-defining term. In the direction genus, it is a word that is too general to be defined (there is no larger class that includes it). In our example that term is "substance," which

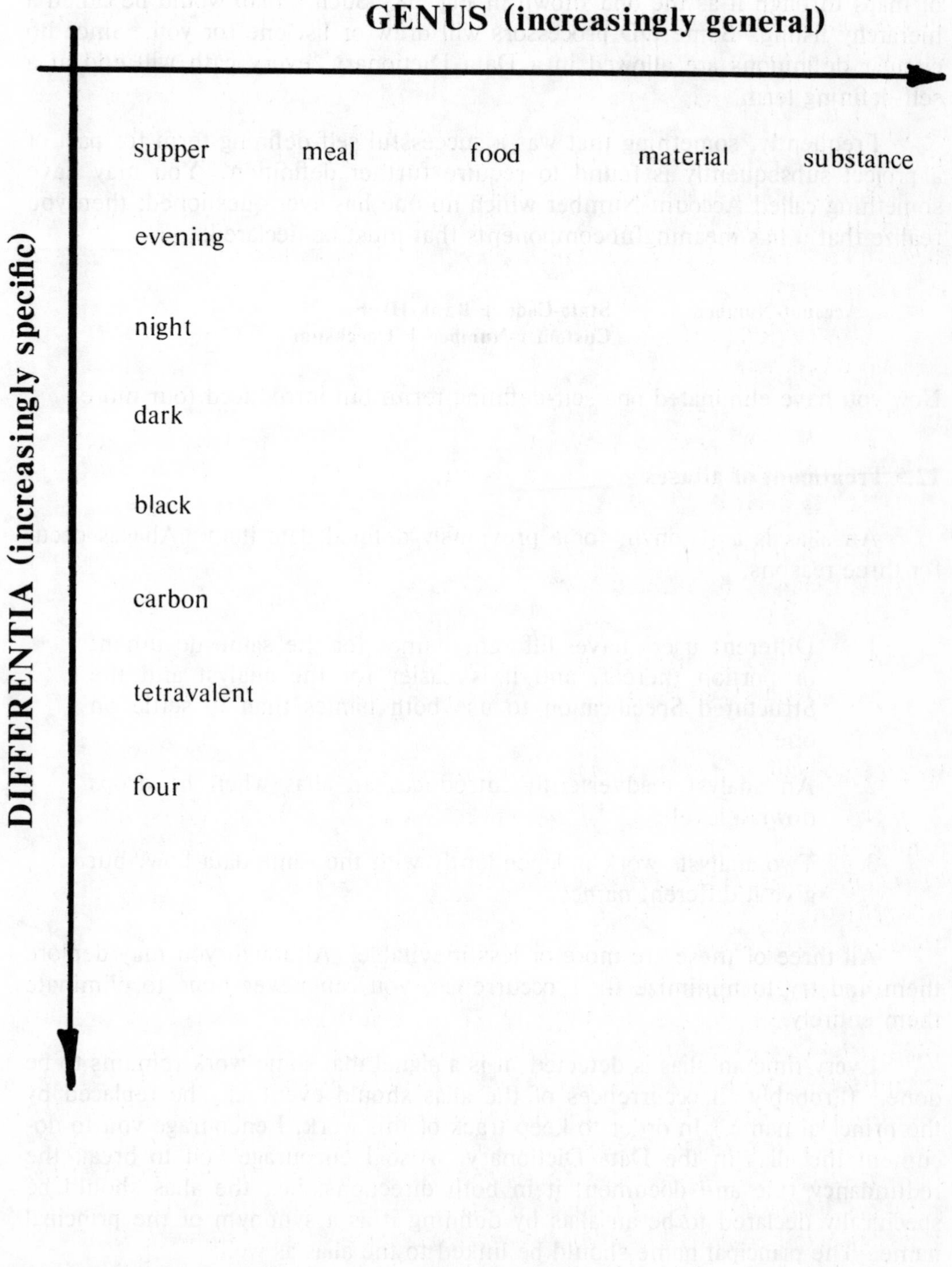

Figure 58

can only be defined circularly by going back to "material." In the other direction is the self-defining term "four," a word too specific to allow further differentiation.

Since a Data Dictionary is a connected set, you could draw the same kind of maps through it as the one shown in Fig. 58. Such a map would be called a hierarchy listing. Some DD processors will draw or list one for you. Since no circular definitions are allowed in a Data Dictionary, every path will end at a self-defining term.

Frequently, something that was a successful self-defining term for part of a project subsequently is found to require further definition. You may have something called Account-Number which no one has ever questioned; then you realize that it has meaningful components that must be declared:

Account-Number   =   State-Code + Bank-ID +
                       Customer-Number + Checksum

Now you have eliminated one self-defining term, but introduced four more.

12.5 Treatment of aliases

An alias is a synonym for a previously defined data item. Aliases occur for three reasons:

1. Different users have different names for the same document or portion thereof, and it is easier for the analyst and the Structured Specification to use both names than to settle on one.

2. An analyst inadvertently introduces an alias when he drops down a level.

3. Two analysts work independently with the same data flow, but give it different names.

All three of these are more or less inevitable. Although you may deplore them and try to minimize their occurrence, you can never hope to eliminate them entirely.

Every time an alias is detected, it is a signal that some work remains to be done. (Probably all occurrences of the alias should eventually be replaced by the principal name.) In order to keep track of this work, I encourage you to document the alias in the Data Dictionary. Also I encourage you to break the redundancy rule and document it in both directions; i.e., the alias should be specifically declared to be an alias by defining it as a synonym of the principal name. The principal name should be linked to the alias as well:

$$\text{Order-for-Passage} \quad = \quad \begin{aligned}&\text{Flight-Number + Date +}\\ &\{\text{Passenger-Name + Fare-Basis}\}\end{aligned}$$

$$= \quad \text{Passage-Order}$$

$$\text{Passage-Order} \quad = \quad \text{Order-for-Passage}$$

Since all but the user-specified aliases will be merged before the end of analysis, this redundancy will eventually be eliminated.

12.6 What's in a name?

"I don't know what you mean by 'glory'," Alice said.

Humpty Dumpty smiled contemptuously. "Of course you don't — till I tell you. I meant 'there's a nice knock-down argument for you!' "

"But 'glory' doesn't mean 'a nice knock-down argument'," Alice objected.

"When I use a word," Humpty Dumpty said, "it means just what I choose it to mean — neither more nor less."

—From Lewis Carroll's *Through the Looking Glass*

Alice is a rationalist, and Humpty Dumpty a nominalist. No matter where you stand on the question of rationalism vs. nominalism, you must be very Humpty-Dumptian about definitions in your Data Dictionary. The names you give your data items should be thought of as monickers, handy ways to refer to them. It is the definition that says what the real composition of the item is, not the name. You might be surprised that Cost-of-Sales, defined as

$$\text{Cost-of-Sales} \quad = \quad \text{Base-Cost + Cost-Allocation-Code}$$

carries some extra baggage with it, something that you do not think of as implied by the name. Perhaps it is misnamed. Nonetheless, its constitution is precisely what the Data Dictionary says it is, neither more nor less. Just as though all our data flows were obliged to carry abstract names like X224 and Zbfrgx.

It is inexcusable for a name to be outright deceptive, such as:

$$\text{Purchaser-Category} \quad = \quad \left| \begin{aligned}&\text{Number-of-Potatoes}\\ &\text{Number-of-Turnips}\\ &\text{Batting-Average}\end{aligned} \right| \quad \text{+ Fudge-Factor}$$

Proper allocation of names provides readability to the top levels. Still, it is the definition, not the name, that matters.

All of the above, of course, does not apply to self-defining terms. For a self-defining term, the name is the only thing that matters.

12.7 Sample entries by class

I said that the definition of a Data Dictionary item would, in general, depend on its class; but then I established defining procedures and operators that did not distinguish among data flows, data elements, and files. In this section, I will go back and show you, using examples, how the entries differ. I will also point out the difference between a definition and an entry — for, in some cases, an entry contains more than just a definition.

Fig. 59 is a sample data flow entry. It consists of four parts: Name, Alias, Composition, and Notes.

The definition proper goes into the Composition section. If there is an intermediate data flow used in the definition, that intermediate might itself be defined in the same entry. Your decision to put it there or in its own entry depends on your estimation of whether or not the intermediate is of any use outside the scope of this definition.

The uses of the Name and Alias sections seem obvious: The entry is going to be accessible by Name; synonymous data flow names are written in the Alias section. Where the entry is for an alias data flow (not the principal name), the Composition section is left blank. In that case, the alias is the definition.

DATAFLOW NAME: INQUIRY

ALIASES: NONE

COMPOSITION:

    [CUSTOMER-STATUS-INQUIRY
      STOCK-INQUIRY
      ORDER-STATUS-INQUIRY
      RAW-MATERIAL-INQUIRY]

NOTES:

    1. ONLY SALES TERMINALS GENERATE INQUIRIES
    2. 20 SEC. RESPONSE TIME

Figure 59

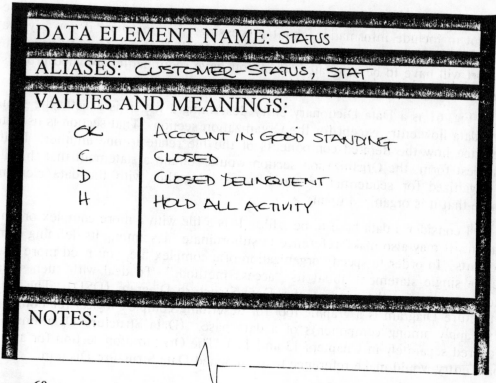

Figure 60

The Notes section is used to record any other information you might like to keep track of on a per-entry basis. There are no rules about Notes except that you must take care not to waste your time recording attributes of the entry that are of no real import, or which are as yet uncertain and meaningless at this early stage of the project. In this last category, I mean to include nearly all information about the physical character of the data flow.

Fig. 60 shows a completed data element entry. Since, by definition, a data element is indivisible, all you can say to define it is what values it may take on and what the meanings of those values are. The table of values and meanings is the definition.

Just as with other data flows, a data element needs a name by which it can be accessed; and it may have an alias. Data element aliases are treated the same as data flow aliases.

There are two kinds of data elements: discrete and continuous. The example shown in Fig. 60 is a discrete one. If it were continuous, it might be defined by expressing valid and invalid ranges of values. In most cases, a continuous data element can safely be left as a self-defining term; i.e., you don't need to define it unless there is some unusual characteristic about its distribution of values.

In specifying meanings of the various values or ranges of values, be careful not to include information about processing. For instance, if you say that the meaning associated with the value OVERDUE is that a Credit-Exception-Report will have to be generated, then you have redundantly specified processing of the data flow that takes on that value.

Fig. 61 is a Data Dictionary entry describing a file; it is exactly the same as a data flow entry except for the Organization section. That section is used to describe how the iterated components of the file relate to one another. In its simplest form, the Organization section would contain a statement that the file is organized for sequential or direct access, together with the data element name that it is organized upon.

I consider a data base to be a file. It is a file with a more complex organization. It may also make reference to subordinate files among its defining components. In order to specify organization of a complex file, you need more than just a single statement about its "access method." To deal with such situations, we require the concept of a Data Structure Diagram (DSD). The Data Structure Diagram is a graphic tool for describing complex logical relationships that apply among components of a data base. (Data Structure Diagrams are covered separately in Chapters 13 and 19.) The Organization section for a data base entry would make reference to an attached Data Structure Diagram.

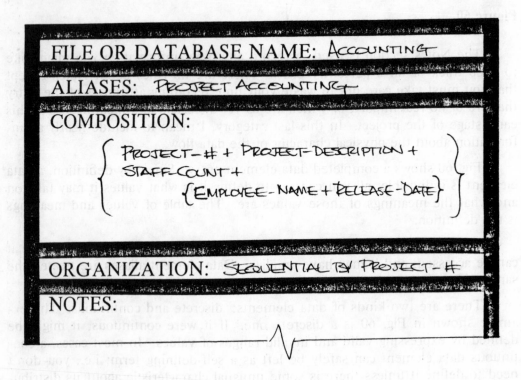

FILE OR DATABASE NAME: ACCOUNTING

ALIASES: PROJECT ACCOUNTING

COMPOSITION:

{ PROJECT-# + PROJECT-DESCRIPTION +
STAFF-COUNT +
{EMPLOYEE-NAME + RELEASE-DATE} }

ORGANIZATION: SEQUENTIAL BY PROJECT-#

NOTES:

Figure 61

Fig. 62 shows a sample Data Dictionary entry for a process. The sections of a process entry are Name, Number, Description, and Notes. The number referred to here is the Bubble Number from the associated DFD primitive.

Our decomposition goal has been to break processes down to the point where each primitive can be described in a single page of Structured English. This may not always be possible. In cases where several pages are required to describe a process, you may want to create some subordinate processes. Each subordinate would have its own process entry in the DD. The subordinate entries would have no number assigned. It is possible that a subordinate process would be used by more than one parent process.

PROCESS NAME: Order Books

PROCESS NUMBER: 1.7.2

PROCESS DESCRIPTION:

1. For each Line-Item on the Coded-Order:

    1.1 Access the Book-Inventory-File, using the Book-Number as key.

    1.2 If the Qty-Available of that book is greater than or equal to the Qty-Ordered

        1.2.1 Set Qty-Sent = Qty-Ordered

    Otherwise (not enough available)

        1.2.2 Set Qty-Sent = Qty-Available

    1.3 Decrement Qty-Available by Qty-Sent

    1.4 Increment Qty-Held by Qty-Sent

2. If the Qty-Sent is zero for all Line-Items (i.e., no books available for any ordered Book-Number)

    2.1 Put entire Book-Order into an envelope

    2.2 Add in a Sorry-Letter

    2.3 Send it all back to the customer

    Otherwise (at least one of the ordered books was available)

    2.4 Write up the Sales-Form

Figure 62

13 LOGICAL DATA STRUCTURES

So far I have spoken of data definition as though it consisted of no more than a description of the *composition* of the item to be defined. That is true for the majority of entries in a Data Dictionary. In particular, once you have described the composition of any data flow or data element, it is completely defined. But what of files? Clearly some additional information about them is required before we can say that their definitions are complete.

Chris Gane and Trish Sarson have written in their new book (cited in the Bibliography) that

"Data flows are data structures in motion; data stores are data structures at rest."

What they refer to as a "data store" I have been calling a file. In our terms, their statement says that files are static sets of data flows. It follows from this that a file definition will require declaration of the component data flows plus some description of the set relationships that apply among the components.

The set relationships of static structures are much more complicated than the ones we have dealt with so far. Consider this: The definition of a data flow shows all the possible ways that the data flow might be composed. *But any single sample of that data flow will consist of a simple concatenation of data elements.* This will not be true of files — a complex file may be a strongly connected network of interrelated data elements. It may contain pointers, links, queues, stacks, or combinations thereof.

Our simple relational operators will not suffice to describe the more complicated kinds of relationships that can apply in modern file (data base) structures. In fact, no formula-type definition will suffice. A formula is inherently one-dimensional (since you *read* it rather than *look* at it). What is required to describe data base structures is a multidimensional representation. Such a representation is a Data Structure Diagram.

Before we look at Logical Data Structure Diagrams in detail, let's try to establish what our true requirements for complex data structures are likely to be, and under what circumstances we need to be concerned about them.

13.1 Data base considerations

Data base, as I shall use the term, is defined as a special case of a file, one in which the components are related to each other by something more than simple concatenation. Common sequential and direct access files are not data bases according to this definition, since their elements (records) are simply stuck together in order, thus making up the whole.

The only reason to introduce additional relationships (pointers and so forth) among the elements of a file is so that you can access the information in different ways. Thus

A data base is any file that can be accessed by a key other than its ordering key.

Any file is either a simple file or a data base. It is a simple file if you can access it one and only one way; otherwise it is a data base.

13.1.1 Redundancy and Modes of Access

A totally unredundant file can be accessed only in the way it is ordered. Each additional kind of access requires added redundancy. A good example of this is the card file in your neighborhood library. When the librarian decided that three kinds of accessing were required (by title, by author, and by subject), there was nothing to do but triplicate the file. A second example is the phone book: two kinds of access (by name and by type of service offered) means that the entire "file" has to be duplicated.

With computerized files, there is the possibility of a very limited kind of redundancy, a pointer. There is also the possibility of replacing stored redundancy in a file with processing redundancy (searching). But there is no way to provide multiple access modes into a file without redundancy of some sort.

A data base processor is a program which provides multiple access modes into a complex file through the use of *controlled redundancy*. The redundancy in the data base is invisible to the user (it is the domain of the processor alone). When information stored in the data base is to be updated, the user makes the change once; any required redundant updating is taken care of by the processor.

13.1.2 The Analyst's Role in Data Base

In many organizations, the analyst is relatively unconcerned with the data base — that is considered the responsibility of the implementation team, just one of the physical details. In other organizations, the analyst is a data base expert. He not only understands the requirements and possibilities in the abstract, but he also knows how to interface directly with the resident data base processor. Each of these possibilities seems extreme to me. I believe that the proper role of the analyst is somewhere in between. He need not be a specialist in the use of any of the data base processing programs (that function belongs to a "data base administrator"), but he ought to understand the possibilities and requirements of data base in such a way as to guide the user toward sensible exploitation of available facilities.

The analyst is the man in the middle between the user and the data base administrator. He has to be able to elicit some logical expression of data base needs from the user, and then convey that information to the data base administrator who can make it happen. In the other direction, he has to translate output of the data base design process into a form which the user can understand, so that the feedback loop will be complete.

In all communications with the user, it is important to work with logical rather than physical considerations. How his requirements are effected in the data base is a technicality from which the user must be shielded. The analyst has to help the user work out his *logical data base requirement* and write it down in some graphic and comprehensible form. That same form should be employed to show the user the results of the data base design.

13.1.3 *What is a Logical Data Base Requirement?*

A user's logical data base requirement is his private model of how static stores of data should be organized. Typically, a user expresses his view of the model with statements such as

"When I tell the system A, I want it to tell me the X associated with that A."

A and X might be data elements or combinations of data elements.

Suppose that a user had expressed three such requirements:

1.   A's yield X's and Y's.

2.   B's yield Y's.

3.   C's yield X's and Z's.

A composite model of this set of requirements is shown in Fig. 63. In data base terms, Fig. 63 is a *subschema*. A subschema is one user's private model of data structure. A *schema* is a global model of data structure that makes all the private models come true. A schema must be, in some sense, the union of all the associated subschemas.

A drawing like the one in Fig. 63, which is used to present a schema or a subschema, is called a Data Structure Diagram. It is a tool for portraying logical data base requirements.

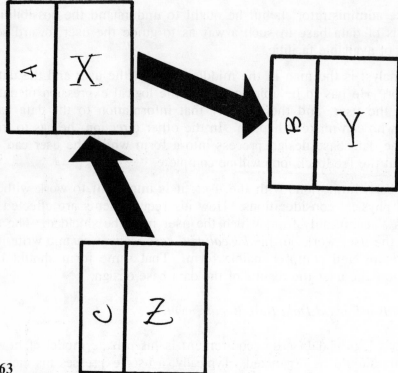

Figure 63

13.2 Data Structure Diagrams (DSD's)

Data Structure Diagrams have been proposed as a means of defining complex files. Without further ado, let's look at an example. Fig. 64 is the defining Data Structure Diagram for something called the Fleet Data Base. The example deals with a shipping application. Fig. 64 shows that the Fleet Data Base is a union of six component files. Each of these is a simple file. (A data base, in this technology, is always made up of simple files, never subordinate data bases.)

To complete the example, definitions of the component files are presented below:[1]

Fleet-File	=	{Ship-Number +
		Ship-Name +
		Planned-Routing +
		Crew-Number +
		Crew-Allocation-Date}
Ship-File	=	{Ship-Name + Ship-Number}

---

[1]The underlined data elements are the accessing keys.

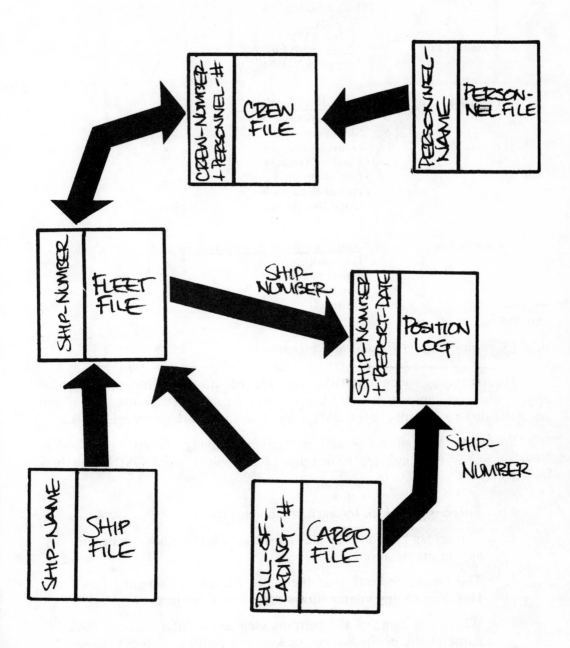

Figure 64

Crew-File	=	{Crew-Number + Personnel-Number + Crew-Member-Name + Seaman-Category + Union-Scale}
Personnel-File	=	{Personnel-Name + Personnel-Number}
Cargo-File	=	{Bill-of-Lading-Number + Customer-Name + Ship-Number + Insured-Value + Loading-Weight + Port-of-Origin + Port-of-Debarkation + Cargo-Description}
Position-Log	=	{Ship-Number + Report-Date + Time-of-Report Report-Source + Longitude + Latitude}

13.2.1 Characteristics of Data Structure Diagrams

Humor me by trying to derive, from Fig. 64 and the definitions presented above, what the convention is for use of Data Structure Diagrams. See if you can write down six points about DSD's and how they work. (No peeking!)

Splendid. Such integrity and inductive capability should be rewarded. Your reward is (you guessed it) my list of six points about DSD's and how they work:

1.   There is one block for each component file.

2.   The access key for each file is written into the block (along the side in my notation).

3.   The arrows indicate that there are pointers linking the files. Direction of the pointer shows direction of the link.

4.   Where the name of the pointing element is different from the name of the pointed-to access key, the pointing element name is written alongside the arrow. Otherwise, it is the same.

5.   Access to the whole involves entering at any block for which you have a key. You then have access to that block plus anything it points to.

6.   There are as many modes of access as there are blocks.

13.2.2 Interpretation of the Sample DSD

The user can access the data base through any of the six keys and examine or modify any element of any subfile he can reach. In particular, he can have an inquiry by Bill-of-Lading-Number that tells him anything about a given order that is maintained in the Cargo-File record. He can also find out the entire history of reported positions for the ship that carries that cargo. He can find out anything about the ship that carries that cargo and its crew (since he has pointers to all that information). He cannot, however, access the Cargo-File entries to find all items on a given ship — Fig. 64 shows no access in this fashion. If he finds he needs access in such a manner (as he almost certainly would), the Data Structure Diagram will have to be changed to suit his revised requirements.

13.2.3 Logical and Physical DSD's

Sometimes a DSD (and hence the data base it describes) will have imbedded physical considerations, such as

- redundant data
- multiple access paths
- imbedded subfiles or repeating elements
- information that is not dependent on the access key contained in the file

To the extent that it does, the DSD will be physical rather than logical. Both physical and logical DSD's are of value in the analysis process. A physical variant may apply to the user's current environment; its logical counterpart will be an important component of the Structured Specification. Deriving a logical equivalent Data Structure Diagram from the current physical description is covered in Chapter 19.

13.3 Uses of the Data Structure Diagram

The analyst makes two rather different uses of the Data Structure Diagram. First, he uses it to write down a description of the current environment in such a way that the user can understand and verify it. Usually, the user's environment — at least from his point of view — can be described in terms of simple files. But, when the user is deeply immersed in data base (i.e., when the current environment is automated and makes use of a data base processor), a Data Structure Diagram will be required.

A second use of the Data Structure Diagram is to help a user reconstruct his true data store requirement, divorced from the arbitrary composition of his current (probably manual) stores. This can be done by the following steps:

- Combine the current files into a data base.

- Draw a Data Structure Diagram for the data base.

- Derive the logical equivalent.

- Decompose back into simple files.

An example of this process will be provided in Chapter 19.

Either of these two uses has the same goal: to include a graphic representation of the user's logical file accessing requirement in the Structured Specification.

14 DATA DICTIONARY IMPLEMENTATION

So much for Data Dictionary in the abstract, its aims and intentions. This chapter presents alternative approaches to effecting Data Dictionary. Its purpose is to answer the question, What set of procedures can we establish to fulfill our Data Dictionary requirement?

The three possible approaches are, again, these:

- automated and manual procedures, centered around the use of a commercial Data Dictionary package

- manual procedures, tailored specifically to analysis requirements of the Data Dictionary

- homegrown automated and manual procedures, usually involving some development effort cr customizing of available software

Each of the three has significant disadvantages: the DD packages now on the market are not oriented toward analysis phase needs; homegrown support is always hard to justify since its cost must be born by a small user group; and manual procedures are so terribly manual.

None of the three directions is ideal, but all are workable. Let's look at them one by one.

14.1 Automated Data Dictionary

Working for even a short time at building and maintaining a manual Data Dictionary is proof enough of the time-consuming and repetitious nature of the tasks involved. If you had your druthers, you would certainly want to have some automated support for this effort, a program to help you manage DD. Such a program would be called a Data Dictionary processor. Before we look at what is available, let's consider what characteristics we would like to see in a Data Dictionary processor.

14.1.1 An Ideal Data Dictionary Processor

For our purposes during analysis, an ideal Data Dictionary processor would be able to do all of the following:

1. Accept definitions as input. Support the four classes of items — data flows, data elements, files, and processes — that we have identified as essential.

2. Provide definition formats and procedures that adhere reasonably closely to the conventions presented in Chapter 12.

3. Allow totally non-redundant input. (If you specify that X is a subordinate of A, you should not be obliged to specify that A is a superordinate of X.)

4. Allow easy updating of definitions.

5. Supply some rudimentary consistency checking. (I include in this category duplicate uses of names, disconnected aliases, circular definitions, syntactically incorrect definitions, and so forth.)

6. Produce, as output, definition listings in alphabetic order by item name. (For ease of maintenance, definitions should be presented in the same format as was used in the original input.)

7. Provide facilities for alias control.

8. Provide some elementary cross-reference listings. (For example, correlations of data elements to their superordinate data flows, listings of data elements required by each process, and listings of undefined terms.)

Our ideal processor would satisfy all these requirements, with a simple enough interface to make it usable by all members of the analysis team without extensive special training.

Along with these bread-and-butter features, there are some bells and whistles that would be handy:

9. Scan the Structured English description of each process and print out an exception list of all data elements used by the process but not shown entering the associated bubble on the Data Flow Diagram. (This is called "lexical checking.")

10. Search for inadvertent sources and sinks of data in the file structure.

11.   Verify balancing of the DFD set at each level.

Since these last few features require that information about connectivity of the DFD's be entered, wouldn't it be nice if the processor could also

12.   Maintain and draw the Data Flow Diagrams.

In fact, there is at least one DD processor (ISDOS) that does draw Data Flow Diagrams. See Section 14.1.3 for a more complete discussion of ISDOS.

Note that unless you are willing to go the whole route of having the DD processor maintain your DFD's, you might be reluctant to ask for features 9, 10, and 11. They would require you to enter a complete connection matrix, showing the input and output data flows for each process. This information is a complete duplicate version of your Data Flow Diagram. Each time you make a change to the DFD, you are obliged to make the corresponding change to the matrix. This redundancy and need for redundant updating seems like a large price to pay for the fancy extras.

An ideal Data Dictionary processor is shown in Fig. 65. Aside from the portion that draws DFD's, it is little more than a special-purpose sort/report generator. Its entire purpose in life is to accept non-redundant input and return it to the user in a number of useful and highly redundant forms. *The ideal Data Dictionary processor provides the advantages of redundancy without the cost of multiple updates.*

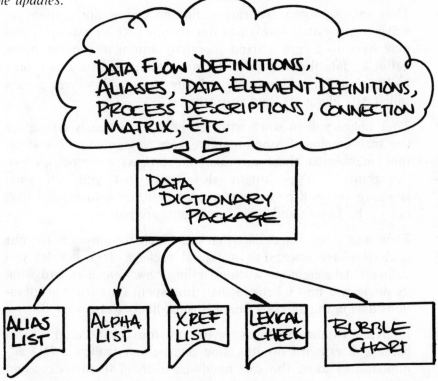

Figure 65

14.1.2 Data Dictionary Packages

There are a number of packages on the market that purport to be Data Dictionary processors. Before you plunk down your $10,000, however, you might be sobered by the reflection that the packages date from the mid and late 1960's, and Structured Analysis didn't come along until 1973-74. Obviously the builders of the packages did not have the particular requirements of Structured Analysis in mind as they went about their work. They weren't very concerned about the analysis phase. Data Dictionary, as they saw it, was a tool for support of the programming process. So the packages they built are oriented around the maintenance and generation of data areas (Data Divisions, COMMON areas, CSECTS and DSECTS, and the like). In addition, most of the currently available Data Dictionary packages are intended to be used with some particular data base processor (IMS, for instance). The approaches and formats of the DD accordingly are closely associated with the approaches and formats of the particular data base.

I do not mean to imply that orientation of the packages toward the later stages of development and toward data base makes them unusable as analysis phase tools. But it does mean that we need to bend our requirement somewhat in order to work with them.

All the DD packages on the market today are, to my mind, subject to these disadvantages:

- They impose some arbitrary definition conventions, many of which seem rather foreign to the analyst. As an example, you may have to accept a fixed hierarchy among the items being defined: data flows may not be subordinate to data flows, only to "transactions"; items subordinate to data flows are always called data elements, and so forth.

- They oblige you to work with terms that are clearly wrong for the analysis phase. Since they are very concerned with coding, they may insist that you call your processes "modules" or "programs." They might also insist that you tell what language you will be coding in, although that information may not yet be known and is, in any event, premature.

- They may insist on details that are much too physical for our analysis phase interests. A typical package does not let you define a data element without telling how long it is and what its mode will be. I believe that time spent on such considerations during the analysis phase is most likely to be wasted.

- Interface to the packages is usually too complicated. Most package users end up allocating one person (called a DD administrator) to be the sole hands-on user. This introduces an additional cost and removes the analysis team from direct control of the Data Dictionary. I have one client who has bought not one, but two DD packages. Because of the people-cost involved in operating either one, he has written them both off and now makes do with a manual scheme.

- The resultant listings are usually out-of-date by whatever the batch turnaround cycle is.

- The initial expenditure in purchase cost plus user training is substantial.

In general, the packages do very well in keeping track of the data characteristics that analysts are little concerned with (formats, lengths, modes, data types, layouts), and do less well with the primary requirement of maintaining definitions.

On the plus side, most packages offer these advantages:

- automated updating

- cross-reference listings galore

- clerical support during the later stages of development, particularly coding

Vendors of DD packages have outdone each other in the number, volume, and splendor of reports available. If you would like to show your boss 11 feet of listings that analyze and cross-correlate your 100-entry Data Dictionary in every imaginable way, then one of the packages is meant for you. As paper generators, they can't be beat.

14.1.3 ISDOS

ISDOS is the name of a project conducted at the University of Michigan by Dr. Daniel Teichroew and his associates. The purpose of the project was to develop automated aids for systems analysis. The resulting system of tools is somewhat more ambitious than a simple DD processor, but there is a DD processor at its heart.

ISDOS allows you to describe data, data movement, and processing in an English-like language called PSL (Problem Statement Language). Your statements are analyzed for completeness and consistency by PSA (Problem Statement Analyzer), which then endeavors to return the information to you in various digested, useful forms. Many of its outputs are graphic. It will draw your Data Flow Diagrams for you if you are willing to live with its conventions.

ISDOS is still not the ideal DD processor, as I have described it. It suffers from its own arbitrariness, initial training cost, and excessive paper generation. But, for the purposes of Structured Analysis, it is probably the best available automated Data Dictionary. It is not at all physical — it is explicitly intended to support the special needs of the analyst. Since there is a host of literature available on ISDOS, I am not going to describe it any further here. One of the key papers about ISDOS is cited in the Bibliography under the name Teichroew.

14.2 Manual Data Dictionary

In my experience, most organizations working with Structured Analysis do not make use of a DD package. They feel that all of the packages, even IS-DOS, impose too many requirements on the analyst and solve too few of the real problems. Many such organizations have established a simple set of manual Data Dictionary procedures that take care of the major needs of definition control. The manual approach means that you get none of the bells and whistles; even a simple cross-reference listing is impossible. The advantage of manual DD is that it allows exact implementation of whatever defining conventions you settle upon, imposing no arbitrary restrictions of its own. While there is a substantial amount of drudge work required to maintain a manual DD, the work is straightforward enough to allow clerical support personnel (e.g., a Development Support Librarian) to handle it with brief training and little supervision.

A workable set of manual Data Dictionary conventions is the following:

1. Create an index card or loose-leaf page for each item to be defined.

2. Write the item name and class on the top.

3. Write the definition using the abbreviated defining operators described in Chapter 12.

4. Use the back of the card or page to write notes about physical characteristics as they become relevant.

5. Make a card for each alias. The definition of the alias is a reference back to the principal name.

6. Make an entry for each self-defining term. (This entry is just a place holder to indicate that there is a self-defining term with that name. Obviously there will be no definition.)

7. Keep entries in alphabetic sequence by name.

14.3 Hybrid Data Dictionary

I term your approach to Data Dictionary hybrid if you start off manually and then augment your procedures using some simple automated tools. You might find, for instance, that you could alleviate the processing drudgery by using a text editor to build and maintain definitions. This might also allow you to hitch up a sort and a report generator to give you some cross-referencing ability. With a very modest amount of work, you may be able to implement something fairly close to the "ideal DD processor" of Fig. 65. We did this at Yourdon. Using the "tinker-toy" utilities of the UNIX operating system, we built a workable DD processor to complement our manual approach. It does not draw DFD's (we still maintain these manually), but it does do all the rest. It allows on-line lookup and change of definitions. Total manpower was all of one work-week.

I know of other organizations that have built Data Dictionary utilities around such processors as

- TSO

- ROSCOE

- WILBUR

- SNOBOL

14.4 Librarian's role in Data Dictionary

Many companies have at last begun to supply some clerical support for the development process. If you work for such an enlightened organization, you may be able to get a librarian assigned to your project during the analysis phase. Such a person could best be used in the area of the Data Dictionary. I offer this list of tasks that an intelligent librarian could assume:

- regular scanning of the DD looking for inconsistencies, loose ends, multiple names, missing entries, and so forth

- deleting of unwanted aliases by replacing each use of the alias name by the principal name

- cross-checking of process descriptions against the DFD's (lexical checking)

- ensuring the integrity of the Data Dictionary, by tracking lost or misfiled entries

- interfacing with the Data Dictionary processor, if one is used, or to any DD utilities

- coordinating of multiple copies of the DD if such exist

Bringing a new librarian up to speed involves the expenditure of less than one work-week of an analyst's time spread over about six weeks. One objection to using librarians is that the job is subject to frequent turnover. Since this is the case, it makes sense to document librarian procedures carefully in the form of a brief self-teaching handbook, complete with examples. You should expect your librarian to move on to bigger and better things within a year or so (perhaps programmer or logical data base aide). Anyone who is intelligent enough to be a good librarian is much too intelligent to be one forever. Fortunately for our purposes, that same observation often can be made about secretaries. Many secretaries are willing to become librarians and serve in that capacity for a year or more because it offers them an improved career path. The kind of methodical dependability that we so often take for granted in secretaries is too valuable to use for typing and copying. Advancing such a person to the position of Development Support Librarian can put those resources directly to work for the analysis team.

14.5 Questions about Data Dictionary

The following are some of the questions that I hear most often when teaching Structured Analysis seminars.

Do users have to work with the Data Dictionary? Yes. Since the DD is an integral component of the Structured Specification, users have to develop at least a reading knowledge of the conventions used. Many users actually become quite facile with the Data Dictionary, even being able to mark up the definitions with corrections in the proper format.

How do you get user acceptance of the Data Dictionary? First of all, the Data Dictionary does pose more of an acceptance problem with the user than do Data Flow Diagrams. Most users are quite familiar with DFD's since they resemble the tools they have been working with all along. But the Data Dictionary is something quite new. It looks a bit like formulas and equations, the kinds of things that tortured so many non-technical people through mandatory courses in algebra and chemistry. You might start by using the spelled-out (capital letter) operators, and then shift to the abbreviated form after the concept is clear. I think it is also important to let the user see you working with DD formulations before he ever realizes that they are going to become part of the specification. Introduce your convention as a little trick you use to keep track of interfaces. Make sure he sees a number of definitions during the early interviews. Let him look at your notes, just to verify them. You have to sell the idea as a useful way to write down the data content of a file or interface without getting all hung up on formats.

Users are used to seeing report layouts included in the specification. How do you deal with this? You cannot include both a Data Dictionary definition and a layout of an item in the specification without introducing unwanted redundancy. I suggest you try to convince the user that the definition is what's really required during the early part of analysis; the formats ought to come later. That allows you to deal with one concept at a time. After you have co-opted him into using DD during the analysis phase, he may be amenable to deferring the actual formatting until implementation. The ideal situation is when the programmer does the formatting and the user is allowed to accept or reject the results. In order to sell him this idea you have to provide a standing offer to redo any format that he doesn't like. If you go about formatting in an intelligent and professional manner, he will not be inclined to overwhelm you with a lot of changes. The advantage is that you do not include prematurely physical information (formats) in your specification, and hence expose yourself to the need for additional modifications later on. The advantage for the user is that he gets to work with reports for a while before he makes his final specification of formats.

Suppose the user insists on having layouts and formats in the specification? Then you have to get rid of the definitions (sob), letting the formats stand in their place. It makes your specification much less maintainable and much more subject to change — obviously, you would want to include the formats as late as possible.

How do you work with Data Structure Diagrams when you have a Data Dictionary package? Some packages (usually the ones that are integrated into data base processors) provide facilities for treating Data Structure Diagrams as schemas and subschemas of the data base.

It sounds like you are not enamored of any of the packages. Would you discourage their use by organizations just getting into Structured Analysis? No. In general, I would encourage organizations to look into the use of one of the packages, but only if it were already up and running and supported in their center. Otherwise, I think introduction of an expensive and complicated piece of support software would be an unwise burden to add to a pilot project using new techniques.

PART 4

PROCESS SPECIFICATION

15 DESCRIPTION OF PRIMITIVES

Structured Analysis would have you spend most of your analysis phase time drawing Data Flow Diagrams and creating the Data Dictionary. I hope you realize that in working with these tools, you will be doing the essential partitioning and interface analysis, but not really *specifying* anything. When you have finished your DFD's, you will have indicated precisely what needs to be specified. Then you have to go back and do it. The real specification work consists of writing concise descriptions of the bottom-level bubbles (functional primitives). It is these descriptions that I have been calling "mini-specs."

Chapters 15 through 17 present techniques and guidelines for writing mini-specs.

15.1 Specification goals

Our set of specifications must fill the same function that the differentia provide in a Data Dictionary definition: to distinguish the system that we want to see built from all other possible systems. I am sure you have seen enough bad attempts at this to be able to define what some of our "non-goals" are:

- confusing description
- ambiguity
- tons of paper
- underspecification
- overspecification

So it follows that a good specification ought to be clear, unambiguous, concise (to avoid excessive volume), and complete (so that no essential element is left unspecified). In order to avoid the sin of overspecification, it should be logical and not physical — it should state what has to be accomplished by the system rather than how the system should accomplish it. In addition to these requirements, we have one other which results from our structured approach to analysis: The specification writing that we do is part of a larger whole (consisting of DFD's, DD's, and so forth), so it must fit in harmoniously with the other components.

Before we can make a complete statement of our goals in writing specifications, we need to decide the following:

- What is the role allocated to a given mini-spec?

- How does it relate to the other elements of the Structured Specification?

- What characteristics determine its quality?

15.1.1 Role of the Mini-Spec

Imagine that you have done a complete and successful partitioning of a system (either the current or the future system) and expressed this in a set of Data Flow Diagrams and an associated Data Dictionary. Although you may have chosen to work with leveled DFD's, they can be thought of as an alternate representation of a single-level diagram like the one shown in Fig. 66. In that figure, we see the whole presented as a network of functional primitives.

Just for the moment, consider the idea of implementing the requirement shown in the figure using a network of connected microcomputers, one microcomputer per bubble. This would be an extreme case of a distributed system. Note that the interfaces among the microcomputers can be read directly from the DFD. (By the way, this line of reasoning is a clue as to why analysts involved in distributed systems are so tuned in to Structured Analysis.) Now, ask yourself what remains to be specified for your distributed system. The answer is that we require one specification document for each microcomputer. That document has to describe what goes on inside the micro, and nothing else (the rest is already done in the DFD and DD).

Specification Goal #1: There must be one mini-spec for each functional primitive in the set of Data Flow Diagrams.

Each micro in our network can be thought of as a transformer, i.e., an automated process to transform its input data flows into its outputs. The mini-spec has to state the rules which govern this transformation.

Specification Goal #2: Each mini-spec must describe rules governing transformation of data flows arriving at the associated primitive into data flows leaving it.

15.1.2 Policy and Method

What are the rules that govern transformation? Let's take an example. Fig. 67 portrays a DFD segment, a primitive together with its input and output data flows. Suppose we ask the user to tell us about that transformation. He is all too likely to say something like this:

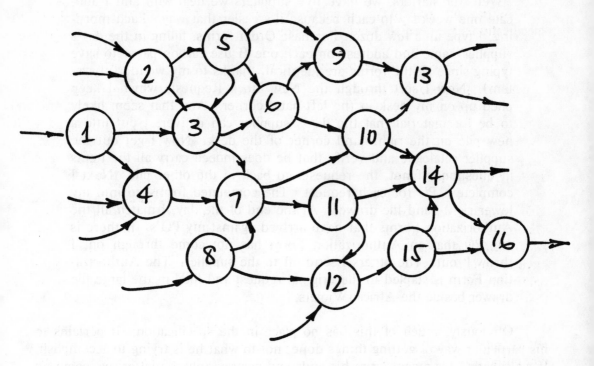

Figure 66

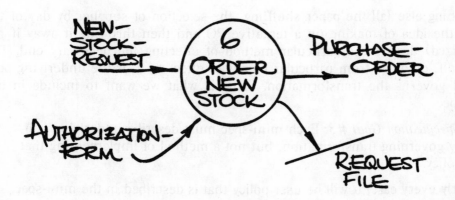

Figure 67

"Well, for starters, we have five suppliers we deal with and I allocate one weekday to each because it's easier that way. Each morning, I type up a few dozen Purchase Order forms, filling in the day's supplier name and address on each one (I use carbon paper to save typing since all the forms are identical, thanks to my wonderful system). Now I sort through the New Stock Requests which I keep piled up on my desk in the left-hand corner. Any that seem likely to be for material that the day's supplier can deliver, I put into a new pile on the right-hand corner of the desk. Now I get out the supplier catalogue and verify that he does indeed carry all the items in question. If not, the requests go back in the other pile. Next I complete a PO for each request. The completed forms go into my lower right-hand file drawer. At the end of the day, I match all the Authorization Forms that have arrived against my PO's. If there is any PO that an Authorization Form has not come through on, I throw it out. The others I send off to the supplier. The Authorization Form is stapled to the original request and filed in the large file drawer beside the African Violets."

Obviously, much of this has no place in the specification. It pertains to his particular way of getting things done, not to what he is trying to accomplish. If we separate his means from his ends and portray only the latter, we come up with something like this:

1. Make up one Purchase Order for each authorized request (New Stock Request with matching Authorization Form).

2. Select a supplier that can deliver the goods requested. Send the PO to him.

3. Save the New Stock Request together with its matching Authorization Form in the Request-File.

Everything else (all the paper shuffling, the selection of supplier by day of the week, the idea of making up a tentative PO and then throwing it away if not authorized) describes a particular method of effecting the necessary end. The end itself, divorced from particular means of effecting it, is the underlying policy that governs the transformation. This is what we want to include in our mini-spec.

Specification Goal #3: Each mini-spec must describe underlying policy governing transformation, but not a method of implementing that policy.

In nearly every case, it will be user policy that is described in the mini-spec.

15.1.3 Relationship to the Rest of the Structured Specification

Consider the three statements that make up user policy for the primitive shown in Fig. 67. Do they constitute a mini-spec? In one sense, they still overspecify the policy for our purposes, because they introduce some redundancy. Can you spot it?

If you look at the statements alone, they seem to be totally non-redundant. But remember that the mini-specs are only part of the Structured Specification. It is the entire result that ought to be non-redundant. By stating where the outputs go (to the supplier or to the Request-File) after they are formed, the description specifies redundantly items that are present elsewhere in the Structured Specification. In this case, it introduces redundancy between itself and the DFD.

> *Specification Goal #4:* The mini-spec must state transformation policy without introducing redundancy of any sort into the Structured Specification.

15.1.4 Orthogonal Representation

To complete our specification goals, I propose to establish some measure of the quality of a given policy description. What is it that distinguishes the good attempts from the bad?

Presented in Fig. 68 for your amusement and amazement is a poor policy description. This one is courtesy of the telephone company — a page ripped out of the San Francisco phone book. If it is not obvious that this is a poor attempt to describe the underlying policy, I suggest that you try to determine from the table what a three-minute call to Venezuela will cost if it is made at 2 A.M. on a Tuesday. Also, try to figure out the meaning of the heading "5 P.M. - 5 A.M. Mon. thru Sat." Does it mean any time between 5 P.M. Monday and 5 A.M. Saturday; or does it mean any time between the hours of 5 P.M. and 5 A.M. on any day but Sunday; or does it mean starting at 5 P.M. on any day but Sunday until 5 A.M. the next morning? Any of these three interpretations could be defended. How about a call that starts at 5 P.M. precisely, or one that starts at midnight between Saturday and Sunday? Look at the line "specifying" rates to the United Kingdom. Would it surprise you to learn that an operator-assisted station call to London costs $3.60 for the first three minutes provided you make it from a phone that is not equipped with IDDD? In fact, that is the case.

Since I am being persnickety about Fig. 68, let me also point out some arbitrary conventions it uses that make it even more difficult to understand:

1.   What is the significance of including the notes under the heading "Overtime Rates?"

international long distance

INTERNATIONAL DIRECT DISTANCE DIALING (IDDD) is available now from certain prefixes in California. Gradually, IDDD will be extended to other prefixes until all telephones are equipped to dial direct to the foreign countries listed below. You will receive a folder of dialing instructions when IDDD is available from your phone. Meanwhile, if you wish to place an international call, dial Operator. If you wish more information about IDDD, please call your business office.

STATION RATES

Typical rates are for the first three minutes and do not include tax.

Charges are based upon rates in effect at the time of connection at the calling point. This also includes calls beginning in one rate period and ending in another.

	5 AM-5 PM Mon. thru Sat.	5 PM-5 AM Mon. thru Sat.	ALL HOURS Sundays	ALL HOURS Every Day
Australia	$ 9.00	$ 9.00	$ 6.75	$ —
Belgium	6.75	5.10	5.10	—
Denmark	6.75	5.10	5.10	—
France	—	—	—	6.75
Germany	6.75	5.10	•	—
Greece	—	—	—	6.75
Hong Kong	—	—	—	8.00
Israel	—	—	—	9.00**
Italy	6.75	5.10	5.10	—
Japan	9.00	9.00	6.75	—
Luxembourg	6.75	5.10	5.10	—
Netherlands	6.75	5.10	5.10	—
Norway	6.75	5.10	5.10	—
Philippines	9.00	9.00	6.75	—
Rep. So. Africa	—	—	—	9.00
Spain	6.75	5.10	5.10	—
Sweden	6.75	5.10	5.10	—
Switzerland	—	—	—	6.75
United Kingdom	5.40	4.05	4.05	3.60 #
Venezuela	†	†	6.50	—

OVERTIME RATES

For each additional minute the charge is 1/3 the rate shown.
• $6.75 from 5 a.m. to 5 p.m. $5.10 from 5 p.m. to 5 a.m.
•• Except Saturday and Sundays when the rate is $6.75 all hours.
† $8.00 from 5 a.m. to 6 p.m.
  $6.50 from 6 p.m. to 5 a.m.
Applies only to calls you dial direct and complete without operator assistance.

© The Pacific Telephone and Telegraph Company 1977.

Rates are those in effect on June 24, 1977. They may change if authorized by the Federal Communications Commission.

19

Figure 68

2.   What is the significance of using white numbers on black background for some rates?

3.   What is the significance of placing table headings in left-leaning, upright, or right-leaning perspective drawn boxes? Why are perspective drawings used at all?

4.   What is the significance of the qualifier "typical" in the first sentence on the left-hand side? Are there some atypical rates?

Of course, the answer to all these questions is that there is no significance whatever to the conventions used. Their only imaginable purpose is to add grace, charm, and artistic distinction to the page. As such, they are totally out of keeping with what the page should be trying to accomplish, i.e., to tell you about international station rates. They simply distract.

What's the matter with the table presented in Fig. 68? See if you can come up with a one-sentence abstraction of its flaws — what do they all have in common? My one-sentence abstraction follows:

The representation chosen for the table is not orthogonal, that is, it does not make explicit provision for all the information to be specified, and not all the information presented is meaningful.

The word orthogonal is taken from mathematics. In its simplest form it means "not overlapping." We often hear the term applied to software languages: "Pascal is more orthogonal than PL/I." A language is considered orthogonal to the extent that it provides the minimum set of constructs needed to write all programs, and it contains no constructs that have any overlapping usefulness with any of the others. A completely orthogonal language would provide one and only one possible way to write any given program. (There are no completely orthogonal languages.) Highly orthogonal languages tend to be spare, uncluttered, small (few constructs), and elegant.

An orthogonal table to specify the rate structure would have one and only one entry pertaining to any call in question. The entry would contain the rate — there would be no notes or exceptions. The entry would be accessed by a unique set of call characteristics, which together describe the call completely. These characteristics would all be meaningful to the table's user.

I'm sure you can imagine such a table. Since the rate seems to depend on four call characteristics (country, time, day, and whether you dial direct) the orthogonal table would be four-dimensional. There are other orthogonal representations of the same subject matter — for instance, sets of two-dimensional orthogonal tables, or sets of statements like this:

Venezuela station rate is determined as follows: If the call is placed on any day from Monday through Saturday, and if the time of origination falls within the period 5 A.M. to 5:59 P.M., then the rate is $8.00. Otherwise, it is $6.50.

Any representation (means of specification) can be judged according to its orthogonality. Take as an example different means of specifying a point on the globe:

1.   Latitude and Longitude

2.   Polar Coordinate and distance from the North Pole

3.   Country, State, County, Elevation, and Distance from the Geometric Center of Muncie, Indiana

The first two representations are orthogonal; the third is not. Consider the first one. There is no overlap between the role allocated to latitude and the role allocated to longitude; they are functionally independent of each other. Each combination of legal longitude and legal latitude defines one and only one point on the globe. There are no illegal combinations or combinations that define no point or multiple points. There is no point on the globe that is not defined by some combination of latitude and longitude. The same observations apply to the second representation. In the last representation, however, each of the descriptive parameters is somewhat dependent on one or more of the others. Their roles in specifying locations overlap. There are some locations that can't be specified at all using that representation. There are some that could be specified using only one of the parameters (the exact center of Muncie, for instance) and for which the others would only introduce inconsistency.

Obviously, the orthogonal representations above are much more useful for point specification than the non-orthogonal one. I will not burden you with any kind of proof that this will always be the case, but I think you can appreciate the value of orthogonality.

Specification Goal #5: Our method of writing mini-specs should be highly orthogonal.

In other words, we should attempt to develop some minimized set of ways to specify, and then use that set exclusively.

I have not stated as a goal that representation techniques used in writing specifications should be entirely orthogonal. That would certainly be nice if it were possible, but it often is not. (If it were always possible to come up with an orthogonal representation for anything that needed specifying, I would go back and rewrite this section to make it totally orthogonal and hence much clearer.) In general, the best we can hope to do is approach this goal.

15.1.5 What is a Mini-Spec?

In summary, a mini-spec is a document that satisfies our specification goals to the greatest extent possible. It is a written description of underlying policy governing transformation of input into output data flows. There is one mini-spec for each primitive in the DFD set. They are written in a highly orthogonal and non-redundant manner.

15.2 Classical specification writing methods

A classical approach to describing logic and policy has been to write text descriptions. You would not have much trouble coming up with examples of perfectly abominable text policy descriptions (see any publication from the IRS as an example), but in most cases the problem stems not from the method itself, but from the lack of partitioning prior to the attempt to specify. I suggest that if you have partitioned sufficiently, you can't get yourself into too much trouble with text specification.

You can do better, though. Inherent weaknesses of the English language (or any natural language) as a tool for specification include the following: It is imprecise, wordy, redundant, and full of implications, connotations, and innuendo. It is probably the ultimate example of a non-orthogonal representation. These very "weaknesses" are what give English its remarkable depth and vitality as a medium for artistic communication. But for specification, they only get in the way.

So what are the alternatives?

15.3 Alternative means of specification

I see only two possibilities for replacing English language text specification: either we adopt a heavily pruned subset of English which is not as subject to the weaknesses signalled above, or we opt for a non-linguistic specification method involving, for instance, tables and lists and graphics. Some policies lend themselves well to linguistic specification, and others do not. I suggest that we follow up on both of these alternatives, and use each one where it applies.

In selecting a subset of English, we want to end up with the most orthogonal possible result. So we would proceed like this:

1.   Throw out every word that has a simpler or more appropriate synonym.

2.   Throw out every sentence syntax that can be replaced by a simpler syntax or combination of simpler ones.

3.   Throw out every way of formulating logic that can be replaced by a simpler way or combination thereof.

     and so forth.

Each time we trim out non-essential facilities of the language, we increase its orthogonality. A subset of the English language which has been restricted in this manner would be called "Structured English." That is what "structured" means when we use the word in a formal way: limited in order to increase orthogonality.

Among the non-linguistic tools for specification, there are many which have stood the test of time. The most important of these, in my opinion, is the Decision Table. A Decision Table is a table of entries that are actions and policies. The table is entered by one or more conditions that apply to a given situation, and the found entry determines the policy for dealing with that situation. A Decision Tree is a graphic presentation of a Decision Table. Decision Trees are sometimes useful for users who are frightened by Decision Tables.

The alternatives to classical text description are the following:

- Structured English
- Decision Tables
- Decision Trees

These are referred to as the structured tools for description of logic and policy.

In general, I encourage you not to try and settle on any one tool and use it exclusively for all mini-specs, but rather to pick and choose among them, selecting the proper tool for each situation as a function of the policy to be described and the ease of communication with the particular user involved.

The same tools used to describe policy during the analysis phase may be used during the later phases to describe processing logic.

16 STRUCTURED ENGLISH

I trace the term Structured English back to a landmark paper by Caine and Gordon called "PDL — a tool for software design." (A complete reference is given in the Bibliography.) To be honest, they were not talking about exactly what analysts have come to call Structured English, but about something we now refer to as pseudocode. The Structured English I advocate here would not be a good tool for software design, since the requirements of design impose too much formalism and "codesyness" for an analysis-level tool. (Still, I think the Caine/Gordon paper is extremely relevant to the subject of linguistic specification; I urge you to read it.)

On the theory that you did not rush off to read the PDL paper in spite of my urging (there is just no respect for authority anymore), I reproduce here part of its definition of PDL:

"It is a *pidgin* language in that it uses the vocabulary of one language (i.e., English) and the overall syntax of another (i.e., a structured programming language)."

That statement could very appropriately be applied to Structured English.

In composing our structured subset of the English language, we will do some extensive trimming of the vocabulary. But the most noticeable difference between Structured English and normal written English is the one cited in the preceding quote: adaptation of the limited constructs of Structured Programming in place of the infinitude of ways that unstructured English allows you to formulate logic. This idea is entirely in keeping with our goal of building an orthogonal tool. The developers of Structured Programming were faced with the exact same task in their work: to trim programming languages by discarding overlapping facilities. The result of their effort is a highly orthogonal set of constructs, readily applicable to our requirement.

16.1 Definition of Structured English

Structured English is plain-vanilla English minus some of the more elaborate facilities of the language. Specifically excluded from our subset are the very features that tend to get analysts into the most trouble while writing specifications:

179

1. wishy-washy qualifiers (adjectives and adverbs)

2. compound sentence structures

3. all modes but imperative (Take heart, friends, this means that you will have to write specifications without using the subjunctive. Should that you but have known!)

4. all but a limited set of conditional and logic statements

5. most punctuation (semicolons, dashes, exclamation points, question marks, ellipses, and the like)

6. out-of-line description, specifically footnotes

All of this is negative — it says what Structured English is not. The following definition describes what Structured English is:

Structured English is a specification language that makes use of a limited vocabulary and a limited syntax. The vocabulary of Structured English consists only of

- imperative English language verbs

- terms defined in the Data Dictionary

- certain reserved words for logic formulation

The syntax of a Structured English statement is limited to these possibilities:

- simple declarative sentence

- closed-end decision construct

- closed-end repetition construct

or combinations of the above.

I suspect that the syntactic limitations may seem a bit obscure to you. Let's look at some examples before we go on.

16.2 An example

Since you are already familiar with the process shown in Fig. 67 and have suffered through the user's description of it, I have chosen that as a subject for an initial example.

===

Policy for Ordering New Stock

For each New-Stock-Request, do the following things:

1.   Search for an Authorization-Form with Reference-Number
       equal to the Request-Number on the New-Stock-Request.

2.   If there is no match, discard this New-Stock-Request.
       Otherwise:

       Write a Purchase-Order for the Ordered-Item.
       From the Supplier-Catalogue, select a Supplier
         that carries the Ordered-Item.
       Copy Supplier-Name-and-Address on to
         Purchase-Order.
       Copy Purchase-Order-Number on to New-Stock-
         Request.
       File New-Stock-Request with Authorization-Form.

===

I shall restrict my comments to the form of the example (if its content needs explaining, the case for Structured English is lost). As you can see, the vocabulary I have used in my Structured English is, for the most part, made up of verbs and terms defined in the Data Dictionary (the hyphenated names), together with some obvious reserved words and phrases like "For each," and "equal to" and "If . . . Otherwise." I have not absolutely denied myself the luxury of an occasional adjective or good old wishy-washy noun — the phrase "following things" in the first statement is a prime departure from the standard — but I easily could have. When I write Structured English, I do it as a two-step process. First, I write the whole policy, adhering rigorously to the standard vocabulary. Then, I go over it again and add some exceptions to improve the appearance. After all, we have to sell this to the user. To see my interim product, go back to the example and cross out every word that is not part of the permitted vocabulary of Structured English. You should end up with something like this:

===

Policy for Ordering New Stock

FOR EACH New-Stock-Request:

1. **Find Authorization-Form SUCH THAT Reference-Number**
    EQUAL TO Request-Number OF New-Stock-Request.

2. **IF NO MATCH, discard New-Stock-Request.**
    OTHERWISE:
       Write Purchase-Order FOR Ordered-Item.
       Select Supplier FOR WHICH Ordered-Item
          appears IN Supplier-Catalogue-Entry.
       Copy Supplier-Name-and-Address ONTO
          Purchase-Order.
       Copy Purchase-Order-Number ONTO New-Stock-
          Request.
       File New-Stock-Request WITH Authorization-Form.

===

(I have capitalized the likely reserved words and phrases in this example to call attention to them.) The result, as you can see, is still precise and comprehensible. But it looks too stilted for our purposes; it looks far too much like code. A little editing makes it a lot more palatable.

I hope you will take a non-religious approach to Structured English and allow yourself to deviate from the standard vocabulary when a clear improvement in readability results. We require this much flexibility in our specification language to make it workable for all cases, and to allow us to tailor it for user acceptability.

So far, I have only commented on the vocabulary used in the example. How about its structure? Does it conform to the limited set of logical constructs allocated to us? In order to determine this, I have drawn a tracing of your stream of consciousness as you read the policy (see Fig. 69). This diagram portrays the logical structure of the Structured English description. In order to make it look as familiar as possible, I have used a flowcharting convention to present it.

As shown in Fig. 69, the overall structure of the sample policy is a closed-end loop. Contained within this loop are a number of imperative statements, and one of these (#2) is a closed-end decision structure with a number of imperative statements nested inside it. Since each portion of the structure is either a simple imperative statement, or a closed-end loop, or a closed-end decision, or some combination thereof, the structure does fall within the standard.

If the last demonstration seemed to involve a certain amount of handwaving, I sympathize. A more methodical approach to verifying the structure of a description will be presented in a later section.

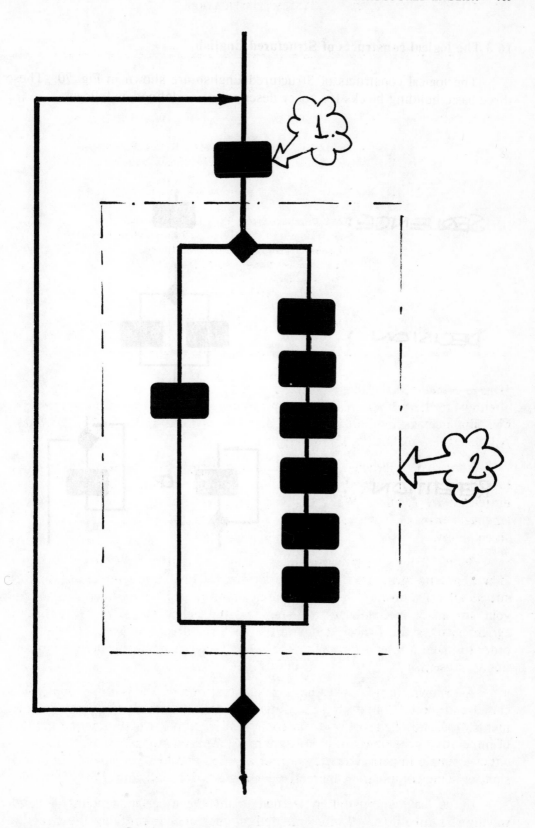

Figure 69

16.3 The logical constructs of Structured English

The logical constructs of Structured English are shown in Fig. 70. These three basic building blocks for policy description are defined as follows:

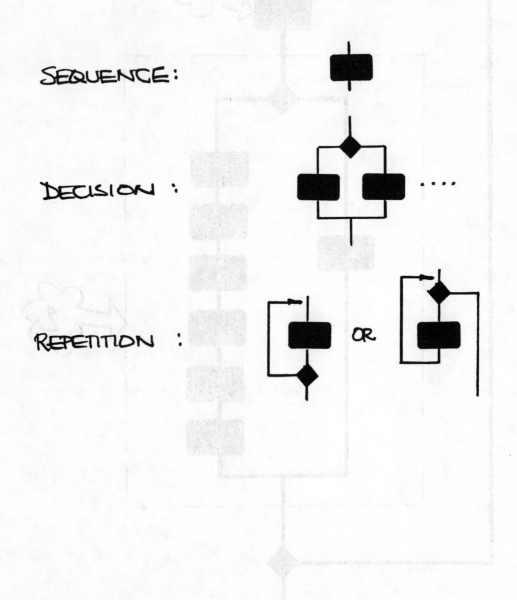

Figure 70

- *The Sequence construct,* consisting of one or more subordinate policies (pieces of the overall policy) which are to be applied one after the other without interruption.

- *The Repetition construct,* a subordinate policy that is done over and over again within some limit.

- *The Decision construct,* consisting of two or more subordinate policies, only one of which applies in any given case.

The three constructs have this characteristic in common: Each has a single starting point and a single ending point. This means that any policy whose syntax conforms to one of the building-block structures will be readable in a simple top-to-bottom fashion, without interruption of the serial thought process. This will also apply to policies that are composed of combinations of the building blocks. The advantage of policies described in this way is that they are easier to read and understand. The center of the brain that is responsible for reading is a serial processor; a description that tries to introduce interruptions in the serial flow of consciousness is difficult to read. (It would be easier to describe such policies with flowcharts. That's why flowcharts were invented.) Policies that lend themselves well to smooth sequential reading are easier to understand. They are not continually fighting against the medium chosen for their representation. I offer no proof that descriptions written explicitly for serial consumption are easier to read, but I hope to have enough examples of both kinds of writing in the following sections to convince you of that idea.

The limited syntax provision that I have imposed upon you means that any properly written Structured English policy description must be *structurally* equivalent to some combination of the three basic building blocks shown in Fig. 70.

But what is the structure of a given description, and how do we determine it? What are we to do about it if it turns out to be different from what we would like to see?

16.3.1 The Structure of a Description

The structure of a policy description is represented by a flowgraph which traces the reader's stream of consciousness as he considers the policy. I am not advocating that you flowchart your policy; that would be unacceptable to most users, and a waste of time. However, just to become familiar with the rather unusual idea of working with a severely limited syntax, I propose that we look at some policies together with their flowgraph structures. You will very quickly dispense with any further need for these graphs as you develop a feel for which policies fit the limited syntax and which do not.

Our first example is taken from the study of a securities exchange. Its associated flowgraph is presented in Fig. 71.

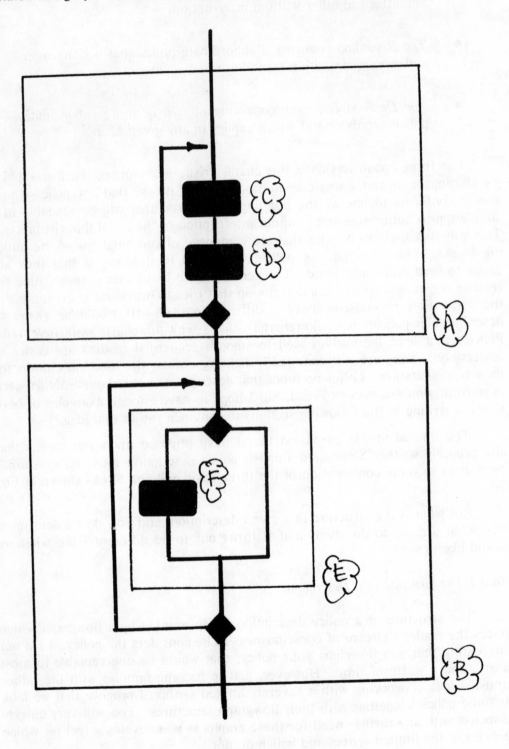

Figure 71

Order Matching Policy Description One

My work involves processing buy and sell tickets contained in two large files (the Buy Book and the Sell Book) and looking for matches. It proceeds like this: First thing each morning, I incorporate any buy orders that have arrived into the Buy Book. Each buy order ticket gets the following treatment: (1) I time-stamp it; (2) I tear off the carbon and pass that to the archivist; (3) I place the ticket in position in the Buy Book by security name, bid price, and time.

Now I go through the Sell Book, starting with the oldest sell order, and check each one against the Buy Book for a match. I consider it a match if these three conditions are fulfilled: same security, same quantity, and bid price within 1/8 of a point of asking price. On a match, I take the two tickets out of the books and staple them together; they now constitute a trade. I give the trade to the data entry clerk.

Does the structure shown in Fig. 71 conform to the limited syntax of Structured English? Yes. The entire policy is single entry, single exit, so it can be viewed as a single Sequence construct. Nested within it are two Repetition constructs (marked A and B in the figure). The first of these (A) has two subordinate Sequences (C and D). The second (B) has a subordinate Decision (E) that itself contains a subordinate Sequence (F). The entire policy and each component of it can be represented in terms of the three building-block constructs.

Since the policy as described by the user already conforms to the restricted syntax, no structural revision will be required to express it in Structured English. All we have to do is trim the vocabulary, incorporate Data Dictionary names, and perhaps indent to emphasize subordination:

==

Order Matching Policy One

For each Buy-Order in Accumulated-Orders-Stack:
     Add Time-of-Day.
     Separate Order-Carbon-Copy.
     Enter into Buy-Book by Security-Name, Bid-Price,
     and Time-of-Day.
For each Sell-Order in Sell-Book:
     Find first Buy-Order with matching Security-Name.
     If found,
          If Qty-Bid equals Qty-Asked and Price-Bid is within
          1/8 point of Price-Asked,
               Combine Buy-Order and Sell-Order into Trade.

==

Before I try to set down procedures for working with description structures, I would like you to look at one more example. This one also involves security order matching:

Order Matching Policy Description Two

I start with the oldest sell ticket in my Sell Book and then proceed through all the rest in the same fashion: First, I search the Buy Book for a match (same security and bid and asked price within 1/8 of a point). If I don't get one, I go back and try the next sell ticket, and so on. If there is a match on security and price, I look at the quantity.

When the quantities are equal, it's easy — I just take out the two tickets and use them to write up the trade. Then I go back to my Sell Book and do the next one in the same way.

When the quantities are different, what I do depends on which is greater.

When the sell quantity is greater than the buy quantity, I rewrite the sell ticket as two tickets: one for the same quantity as in the buy order (I treat this new sell and the matching buy as a trade, and proceed with it exactly as above), the second for the residual quantity. As soon as the trade for the matched pair is complete, I try to match the residual sell instead of taking the next sell ticket from the book. The matching process works the same as above aside from that difference. I may even break up the ticket further. When the sell quantity is less than the buy quantity, I can't break up the buy the way I did with the sell. So I take the two tickets out of the book and pass them on to the floor trader who treats them as a negotiated order.

The flowgraph for this description is shown in Fig. 72. Does the structure of the description conform? No, there is no way to represent the flowgraph of Fig. 72 in terms of the basic building blocks. There are some rather different constructs used in this description, and they are not such handy ones. They don't all have the single-entry, single-exit characteristic.

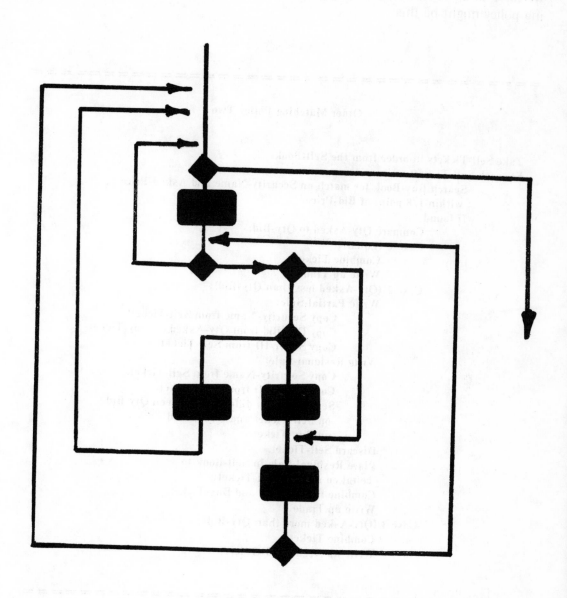

Figure 72

In order to come up with a Structured English description of this policy, we have to do some restructuring. A restructured version of the same underlying policy might be this:

==

Order Matching Policy Two

Take Sell-Tickets in order from the Sell-Book.
For each Sell-Ticket:
    Search Buy-Book for match on Security-Name and Asked-Price
    within 1/8 point of Bid-Price.
    If found,
        Compare Qty-Asked to Qty-Bid:
        Case 1 (Equal):
            Combine Tickets.
            Write up Trade.
        Case 2 (Qty-Asked less than Qty-Bid):
            Write Partial-Sale:
                Copy Security-Name from Sell-Ticket.
                Copy Qty-Bid from Qty-Asked on Buy-Ticket.
                Copy Seller-ID from Sell-Ticket.
            Write Residual-Sale:
                Copy Security-Name from Sell-Ticket.
                Copy Seller-ID from Sell-Ticket.
                Set Qty-Bid to difference between Qty-Bid
                on Sell-Ticket and Qty-Asked
                on Buy-Ticket.
            Discard Sell-Ticket.
            Place Residual-Sale in Sell-Book to
            be taken as next Sell-Ticket.
            Combine Partial-Sale and Buy-Ticket.
            Write up Trade.
        Case 3 (Qty-Asked more than Qty-Bid):
            Combine Tickets.
            Write up Negotiated-Order.

==

The revised description does conform to the restricted syntax (see Fig. 73). Note that it is not the underlying policy which has changed, only our description of it.

16.3.2 An Easy Algorithm for Testing Syntax

If you are or have been a programmer, it will not be difficult for you to look at a flowgraph depicting the structure of a given description and tell at a glance whether or not its syntax conforms. Chances are you will attack it in the same "outside-in" fashion that I used to analyze Fig. 71; i.e., you check that the overall policy at the highest level can be viewed as one or more of the basic constructs. This helps you to isolate the subordinate policies. Then you look inside each of them and see whether they are made up of legal constructs of their subordinates, and so on. If any test fails, the syntax does not conform.

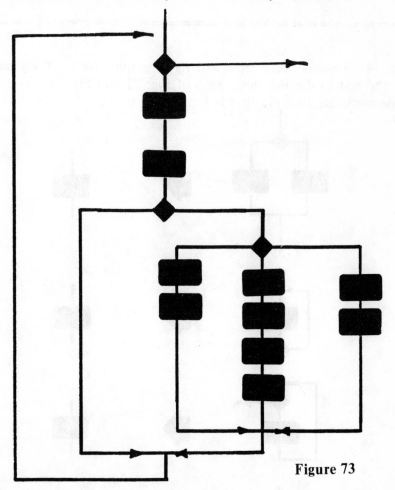

Figure 73

If the graph is complicated, you may have trouble with the outside-in approach. Here is an alternative method (an inside-out method) that is a straightforward and foolproof test:

===================================

Syntax Checking Algorithm

1. **Draw a flowgraph for the description.**
2. **Repeat the following replacement process until no further replacements are possible:**
   2.1 If there is a complete Decision construct,
      Replace it with a Sequence construct.
   2.2 If there is a complete Repetition construct,
      Replace it with a Sequence construct.
   2.3 If there are two adjacent Sequence constructs,
      Replace them with a single Sequence construct.
3. **If the entire flowgraph has now been replaced with a single Sequence construct,**
      The description does conform to the limited syntax.

===================================

Fig. 74 presents these replacements in graphic form. Try your hand by applying the test to the two flowgraphs of Fig. 71 and Fig. 72. You should be able to demonstrate that the first is legal and the second is not. As a further

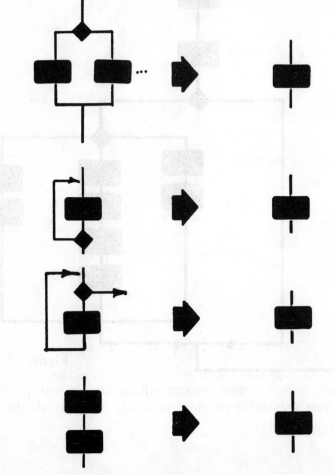

Figure 74

reinforcement of this concept, Fig. 75 presents a set of flowgraphs that conform to the limitations of Structured English, and Fig. 76 presents some that do not.

If you find yourself struggling to determine whether the limited syntax of Structured English has been observed, you may be working with too large a primitive. Consider breaking it down further.

16.3.3 Use of the Structured English Building Blocks

This section will show you some examples of Structured English policy segments that conform to each of the building-block constructs. Each example constitutes a "linguistic implementation" of one of the simple syntactic elements shown in Fig. 70. As we go along, the rules governing these linguistic implementations will become evident.

The first example is a Sequence construct:

==

Policy for Writing up Sales

Access Price-File by Part-Number from the Incoming-Order.
Copy Unit-Price from Price-File-Record into Unit-Price
            of Sales-Form.
Set Subtotal to product of Unit-Price and Quantity-Ordered.
Set Sales-Tax to 3% of Sub-Total.
Set Total to Sub-Total plus Sales-Tax.

==

Alternately, you might consider this example to be five Sequence constructs. Or you might think of it as a single Sequence construct with five subordinate Sequences on the inside. In this last case, you have an opportunity to introduce added clarity by using one statement to declare the whole, and presenting the subordinates that make up that whole in some way that calls attention to their subordinate role:

==

Policy for Writing up Sales

Use Incoming-Order to prepare Sales-Form:
            Access Price-File by Part-Number from the Incoming-Order.
            Copy Unit-Price from Price-File-Record into Unit-Price
                        of Sales-Form.
            Set Subtotal to product of Unit-Price and Quantity-Ordered.
            Set Sales-Tax to 3% of Sub-Total.
            Set Total to Sub-Total plus Sales-Tax.

==

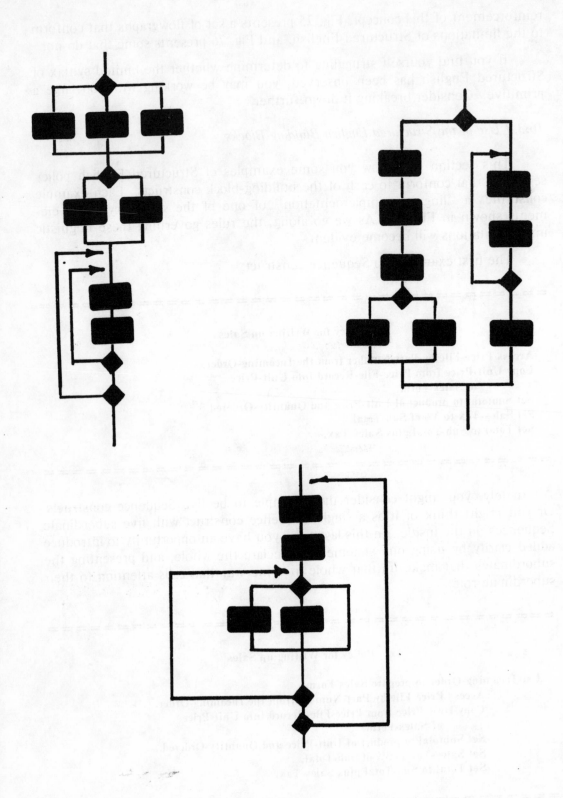

Figure 75

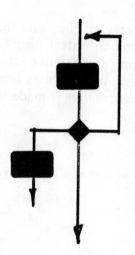

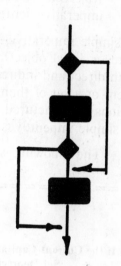

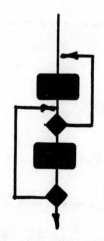

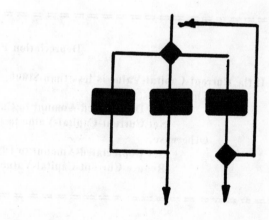

Figure 76

I'm sure you can see the pattern in these Sequence constructs. Abstraction of this pattern gives us our first rule:

Rule One: A Sequence construct is composed of one or more simple imperative sentences.

A simple imperative sentence consists of a single transitive verb (i.e., one that takes a direct object) in imperative mode (as though giving an order) together with direct and indirect object(s). Whenever you use a simple imperative statement or a set of them, you are guaranteed to be conforming to the syntax limitations of Structured English. Most Structured English description is made up of simple imperatives.

The following is an example of a Decision construct:

==

Depreciation Policy

If the Current-Capital-Value is less than $1000,
      Set Depreciated-Amount to Current-Capital-Value.
      Set Current-Capital-Value to zero.
Otherwise,
      Set Depreciated-Amount to 10% of Current-Capital-Value.
      Reduce Current-Capital-Value by 10%.

==

Some analysts would write this in a slightly different format:

==

Depreciation Policy

If the Current-Capital-Value is less than $1000,
      Then,
            Set Depreciated-Amount to Current-Capital-Value.
            Set Current-Capital-Value to zero.
      Otherwise,
            Set Depreciated-Amount to 10% of Current-Capital-Value.
            Reduce Current-Capital-Value by 10%.

==

This format calls your attention to the idea that there is one construct here, not two.

Frequently, one of the legs of the Decision construct involves no action; that is still legally within the syntax of the Decision construct. It would be written like this:

===

Young Driver Surcharge Policy

If Policy-Holder-Age is less than 25,
    Add 12% to Amount-of-Policy.
    Add Insulting-Disclaimer to Notes-Section.

===

From this and the previous examples, you have probably already deduced the following:

> *Rule Two:* A decision construct may be implemented by a conditional statement (typically something like "If suchandsuch is true"), followed by a subordinate policy which applies if the condition is met, and another which applies if it is not. The second of these — the "otherwise" policy — may be omitted.

Most analysts use the If-Then-Otherwise format to build their Decision blocks. The If-Then-Otherwise works like this:

===

If-Then-Otherwise Format

If <fill in your condition here>,
    Then,
        <fill in your "then" policy here>.
    Otherwise,
        <fill in your "otherwise" policy here>.

===

Any description you write using the If-Then-Otherwise format is bound to conform to the syntax limitations of Structured English, provided that you have no violations in any of the subordinate policies. Of course, you may use some different phraseology to express the same logic.

A word about conditions: You will often want to use some compound conditions (If this and that and theotherthing) to distinguish among your subordinate policies. This is perfectly all right providing you don't get carried away. Readability is the key. If the compound condition would be comprehensible in spoken English, then it is acceptable; otherwise, you would be better advised to express part of the condition with a subordinate Decision construct. In any event, never allow yourself to use compounds that are combined with both and's and or's (If this and that or theother) since they make no sense.

Also, be careful that the end of your Otherwise-policy is clearly distinguished from the next construct. You can take care of this with your indentation convention if you apply it rigorously. You can get into trouble with Decision constructs that are spread across the page boundary; in that situation, the reader will lose track of indentation. Some people like to add an end marker to the If-Then-Otherwise format in order to take care of this problem:

===

If-Then-Otherwise Format

If <condition>,
    Then,
           <then-policy>.
    Otherwise,
           <otherwise-policy>.
In any case,

===

Suppose you have a policy that has more than two alternate subordinate policies, only one of which applies. The If-Then-Otherwise format only allows for a two-way Decision construct. Of course, you could use subordinate Decisions to take care of further subdivisions, but that might be artificial, particularly when the situation cries out to be treated with a certain amount of parallelism. A better approach would be something like this:

===

Policy Governing Mode of Shipment

Select the policy which applies:
    Case 1 (Cost-of-Order > $100):
        Send via Air-Freight-Prepaid.
    Case 2 (Cost-of-Order in range $10 to $100):
        Send via Parcel-Post-Prepaid.
    Case 3 (Cost-of-Order < $10):
        Send via Parcel-Post-Collect.

===

The word "case" in English means precisely what we mean to imply here, i.e., the existence of n possibilities, only one of which is applicable. Just as in our normal use of the word in spoken communication, this use of it is single-entry, single-exit. The reader understands instinctively that once he has applied the one case that is relevant, the next portion of the policy is taken from whatever follows the set of cases.

The format used in the preceding example might be called the Select-Case format. In the abstract, it looks like this:

==

Select-Case Format

Select the policy which applies:
      Case 1 (<put condition that defines Case 1 here>):
          <put policy which applies for Case 1 here>.
       .
       .
       .
      Case n (<put condition that defines Case n here>):
          <put policy that applies for Case n here>.

==

This gives us a second implementation of the Decision construct. The rule for it is as follows:

Rule Three: A Decision construct may be implemented using the Select-Case format or its equivalent.

Any policy described using the Select-Case format or something equivalent to it is guaranteed to conform to the syntax limitations of Structured English, provided that the subordinate policies (under the cases) also conform.

It only takes one more example to round out our set. This is an implementation of the Repetition construct:

==

Daily Ledger Policy

For each Passenger-Record in the Reservation-File:
      Accumulate Amount-Due into the Running-Total.
      Build a New-Debit-Record.
      Write the New-Debit-Record to the Daily-Ledger.

==

Depending on the situation and your personal preference, you might choose to describe conditions that govern continued repetition in different ways. For instance, many analysts use catchphrases like "As long as there are any remaining whatevers, continue doing the following" or "Do the following once for each whatever" or some equivalent. Regardless of phraseology, they all more or less fit into what I call the Continue format:

===

Continue Format

<catchphrase and condition for continued repetition here>:
        <repeated subordinate policy here>.

===

Some people like to put the condition for continued repetition at the end:

===

Continue Format

<catchphrase>:
        <repeated subordinate policy>.
<condition for continued repetition>.

===

I still consider this to be in the Continue format. Here is an example:

===

Priority Message Policy

Repeat the following:
        Take the next Priority-Message from the Message-Stack.
        Write Message-Text to Message-Log.
        Write Message-Header and Log-Number to Priority-Queue.
Until there are no more Priority-Messages.

===

The rule that summarizes this is the following:

> *Rule Four:* The Repetition construct can be implemented with a condition that governs continued repetition followed by the subordinate policy to be repeated.

Any description that adheres to this rule — for instance, one that makes use of the Continue format or its equivalent — is guaranteed to conform to Structured English syntax restrictions, provided that any subordinate policies apply.

The examples in this section each have used only one building block. Obviously, Structured English policies need not be limited in this respect. It is perfectly reasonable to have a single policy that makes use of more than one of the building blocks, perhaps all of them. Such a policy would be constructed of concatenated and nested subordinate policies. Here is an example:

===

Bounced Check Retribution Policy

Select the case which applies:
       Case 1 (No Bounced-Checks in Customer-Record):
           Write Exemplary-Customer-Citation to Annual-Summary.
       Case 2 (One Bounced-Check):
           If Yearly-Average-Balance exceeds $1000,
               Remove Bounced-Check from Customer-Record.
           Otherwise,
               Reduce Credit-Limit by 10%.
       Case 3 (Multiple Bounced-Checks):
           For each Bounced-Check,
               Reduce Credit-Limit by 15%.
           Set Credit-Rating to Deadbeat.
           Write Scathing-Comment to Annual-Summary.
           Write Customer-Name-and-Address to IRS-Enemies-List.

===

A final rule which applies to formulation of Structured English policy descriptions is this:

Rule Five: A description which combines syntactically valid subordinate policies is itself syntactically valid as long as the subordinates are related by one or more of the building-block constructs.

16.3.4 Structured English Composition

You can write Structured English by starting out with any description, and then applying these steps:

- Flowgraph the description.

- Untangle the flowgraph.

- Apply the same untangling operation to the policy description.

There is even a documented technique (the Ashcroft-Manna algorithm) to help you do the untangling.

You *could* go about it that way, but I propose a far easier approach — Structured English composition. It works like this:

    1.   Select one or more linguistic implementations for each of the basic constructs. You might settle on phraseology used in this section, or develop something of your own.

2.   Use the catchwords and catchphrases of your implementations to make up a list of reserved words.

3.   Add to your list new words that are necessary for describing conditions, relationships, and so forth, as you detect a need for them.

4.   Restrict yourself in writing policy descriptions to the use of words in your reserved list, Data Dictionary terms, and imperative verbs.

If you go about policy description in this way, you will be unable to write anything but legal Structured English. It's possible that the result may seem a bit stilted and stuffy — if so, go back and relax your vocabulary restrictions to smooth it out. This editing should not affect syntax.

16.4 The vocabulary of Structured English

Why is it that Structured English allows you to use any English language verb, but not any noun or adjective? The reason is that verbs simply cannot get you into the kind of trouble that the other words can. More than a century ago, no less an expert on words than Humpty Dumpty made this point:

"They've a temper, some of them — particularly verbs: they're the proudest — adjectives you can do anything with, but not verbs."

— Through the Looking Glass

A word "you can do anything with" is obviously not going to be much help in writing specifications. Its great flexibility will only encourage varying interpretations — just what we want to avoid. Verbs have more intrinsic meaning; that is what makes them prouder (more independent) in Humpty Dumpty's estimation.

Of course, there are some verbs that aren't very proud. The worst of these are the verbs "process" and "handle." While a statement like "Process New-Customer-Update" seemingly qualifies as valid Structured English, it doesn't have any meaning. "Process" is a catchall verb that implies the union of all verbs that could ever be associated with the given object. A rough translation of the statement "Process New-Customer-Update" is the following:

Do any and/or all of the kinds of things that anyone might ever think of doing to a New-Customer-Update without provoking gales of laughter.

"Handle" is equally bad. I suggest you banish both of these words from your vocabulary. Words like "edit" and "verify" are weak. You should complement their use with specific edit and verification criteria.

If you had to build a formal specification language, you would have to pick and choose among the adverbs and adjectives of English in an attempt to come up with a minimal set of the strongest words. You might accept words like "zero" and "non-zero," and reject words like "peachy." For our purposes, however, this has already been done. When we wrote the Data Dictionary definitions for our data elements, we considered all the adjectives that were meaningful to us. These are the value names we used in the definitions. If we have defined the elements completely and successfully, the set of value names will be entirely sufficient for our purposes. We need not be burdened with a complete general-purpose set of modifiers. The value names constitute a highly orthogonal special-purpose set, tailored to our specific requirements.

Similarly, as part of the process of writing the Data Dictionary, we have already selected our required set of nouns. These are the names given to files, data flows, and data elements. If our Data Dictionary is complete, we shall need no others. All objects (direct and indirect) of Structured English statements will be Data Dictionary-defined names.

In summary, the Structured English vocabulary is made up as follows:

- *Verbs* that can be selected from the set of English language transitive verbs minus blatantly meaningless ones such as "process" and "handle."

- *Objects* that must be taken from the set of names given to files, data flows, and data elements in the Data Dictionary.

- *Qualifiers* that can be taken from the set of names given to data element values in the Data Dictionary.

- *Conjunctions* (such as "if," "while," "until") that must be taken from the set of reserved words used to effect the three syntactic constructs.

- *Relational attributes* (such as "equal," "and," "or") that must be taken from the set of reserved words.

No other words are considered part of the vocabulary of Structured English.

16.5 Structured English styles

When I teach the subject of Structured English and its use, I often get questions like these:

- What is the list of reserved words?
- Can you give us a formal language description?

- Is there a program that will process Structured English and verify its correctness?

- Is there a program that will take Structured English as input and generate code?

I don't have answers for such questions. They all imply a degree of formalism that Structured English simply does not possess. It should be clear from my examples that Structured English is not a rigorous specification language that allows you, effectively, to *code* your requirement. If it were, it would be useless. No analyst is ready for such rigor so early in the project. The last thing we want to do is start coding during the analysis phase.

In the examples presented above, I have not even tried to be consistent in my formats and conventions. On the contrary, I have endeavored to show you some of the varied approaches that different analysts take to writing Structured English. Some like to use a strict indentation convention — others do not indent at all. Some use military numbering schemes or outline format. Still others draw boxes around their building blocks (see Fig. 77 for an example) to call the reader's attention to the logical structure.

POLICY FOR PENDING ACCOUNTS

```
Repeat the following:

    Access the Customer-Account-Record.

    If Account-Status is Delinquent and Balance is under $1,

        Set Account-Status to Pending.
        Accumulate Account-Balance into Total-Pending.
        Set Date-of-Last-Transaction to Today's-Date.

Until there are no more Customer-Account-Records.
```

Figure 77

Your particular approach to Structured English is a matter of style. There is one and only one rule which applies: make it comprehensible and palatable to your user. Some users are quite content to look at descriptions that border on pseudocode. Others will only look at descriptions that seem to be written in normal text. There is a Structured English style to suit each of these extremes.

In selecting a style for use in your own environment, you need to take account of considerations such as:

1.   How terse can the descriptions be without scaring off users?

2.   What limits and conventions shall apply to nested subpolicies?

3.   How can confusion about block boundaries be avoided?

In dealing with these questions, the chief concern must be user characteristics: How much is the user willing to bend toward your methods? How much do you have to disguise your methods to obtain his acceptance?

16.5.1 Abbreviated and Narrative Styles

It is a very rare user who is willing to look at program listings to verify that his requirement is being met. If you are a beginner, this may come as a shock to you. You might assume that users would want to see the actual code, particularly if it were written in some fairly conceptual language such as PL/I or COBOL. After all, that code is the ultimate specification. What a boon to the development process if the user would treat it as a specification rather than insisting on some sugar-coated documentary description.

That line of reasoning is somewhat naive for two reasons: Our current compiler languages are not nearly conceptual enough for true specification work, and the coding process is too time-consuming to suit the iterative exchange of ideas that must take place during the analysis phase. More importantly, the thought that the user might be willing to deal with the code does not take into account the kind of person he is: The typical user is far removed from the world of systems development. Anything that happens in the innards of a computer is likely to be too technical for his peace of mind. Code, no matter what variant of it is used, is a foreign language to him. From his point of view, asking him to read PL/I is as strange as asking him to read Chinese. He can see no benefit in struggling with a foreign language — he is likely to get himself into trouble by agreeing to something that may have disastrous implications and that is beyond his understanding. There is also the question of territoriality: Why should he go so far into your turf, rather than being met on neutral ground?

For all these reasons, users are likely to be turned off to Structured English to the extent that it looks like code or seems to be a tool of system developers. I'll come back later to the question of how you present the concept to your user. In this section we will consider only how the appearance of our Structured English can be modified to suit the user.

The most abbreviated style of Structured English is what you come up with if you strictly adhere to the vocabulary limitations. If you go about composition with the two-step process I introduced earlier, the abbreviated style is your interim product. Here is an example:

===

Policy for Deletion of Suspended Accounts

FOR EACH Customer-Record IN Account-File:
    IF Account-Status EQUAL Suspended,
        IF Todays-Date MINUS Date-of-Last-Transaction EXCEEDS 30,
        THEN,
            Set Account-Status EQUAL Retired.
            Set Date-of-Last-Transaction EQUAL Todays-Date
            Write Customer-Record to History-Log.
            Delete Customer-Record.

===

Again, I have capitalized the reserved words and phrases to help keep track of them.

Of course, your user *could* understand this. The question is, will he be willing to work with it? If you can get him to accept this kind of specification writing, you will save yourself the extra task of revising it to suit him.

If your user is put off by the appearance of the abbreviated format (or if you think he would be), some editing is required. Editing for user acceptance may involve the following:

- relaxing vocabulary restrictions to add words and phrases that have the effect of smoothing out the description

- abandoning capitalization and hyphenation conventions

- Revising indentation conventions

The most extreme editing would be for a user who obstinately refused to deal with anything that looked the least bit foreign. Such a user would require you to *disguise* your Structured English so that it looked just like narrative text. As an example, I have taken the same policy you just looked at in the abbreviated format, and disguised it. Here it is, masquerading as innocent text description:

==

Policy for Deletion of Suspended Accounts

We go through the account file, one customer record at a time. For each customer record, we do these things: First, we ask if it has been marked "Suspended" in the account-status indicator. If so, we ask if the date of the last transaction is more than 30 days before today's date. If that is also true, then we do the following four things: 1. Mark the account status "Retired." 2. Set the date of last transaction to today's date. 3. Write a copy of the customer record to the history log. 4. Delete the customer record from the account file.

==

This is still Structured English. Its syntax conforms rigorously to the building-block constructs. Each sentence object is a Data Dictionary name (although the Data Dictionary entry may be capitalized and hyphenated). All modifiers are data element values. True, there is some additional verbiage, but that should not get in anyone's way. You could reconstruct the abbreviated format easily enough by crossing out each word that is not part of the restricted vocabulary, capitalizing each word that is in the list of reserved words, and rearranging the indentation.

Clearly, there are Structured English styles which lie somewhere between the two extremes. Many analysts, for instance, edit to smooth out the description but retain the indentation. I leave it to you to select the middle ground which best suits your environment.

16.5.2 Nesting Subordinate Policies

The human mind deals rather well with the concept of nesting (your mind is doing that right now by reading this parenthetic comment) provided that certain common-sense rules are observed. You can totally frustrate the communication process by nesting too deeply or too often or over too large a span.

You would probably be willing to accept that comment on face value, but I offer an example (an example that will almost certainly (don't ask me how I know) leave you gasping (figuratively) (not literally) for relief (or at least for the end of the example (which, however (and I use the word ("however") advisedly) will go on and on until you finally collapse exhausted)), relief which could only come in the form of a final (and definitive) (but not immediately forthcoming) end to this paragraph (which ought to have served its purpose (whatever that was!) by this time)), anyway.

In Structured English, you can expect to get away with rather more nesting than you would in text, provided that you adopt a helpful indentation convention to guide your reader (see next subsection). However, there are still limits. I offer you my own observation about nesting: two levels is workable, three is marginal, four or more gives you an incomprehensible product. If you encounter a policy that seems to demand more than three levels of nesting, I suggest you try to decompose the primitive — you have stopped your partitioning too soon. Alternatively, you might consider using a Decision Table or Decision Tree instead of Structured English.

If your indentation convention causes you to indent for each subordinate level, you will end up with a self-policing standard. By the time you have gone down three levels, there is no inclination to indent further because there wouldn't be enough space left on the line to write anything.

16.5.3 Blocking and Indentation Conventions

In most of the examples I have shown you, there was some obvious relevance to the placement of statements on the page. My approach to this "laying out" of Structured English policy description has two goals:

- to delineate the boundaries of adjacent logic blocks
- to show subordination

The reason for the first of these goals may not be clear until you see an example in which a boundary is obscured:

==

An Ambiguous Policy

Access the Customer-Account-Record.
If Account-Status is Delinquent,
Write Account-Number to Officer-Action-Report.
Otherwise,
If Account-Balance is less than $1,
Set Account-Status to Pending.
Accumulate Account-Balance into Running-Total.

==

The problem here arises in the last line. It is not clear whether the accumulation of Running-Total applies only to those accounts with a balance of less than $1, only to accounts which are not delinquent (without regard to balance), or to all accounts. Any of the three interpretations is defensible. In order to decide among them, you need to know where the two decision blocks end. It may belabor the point, but let me show you all three possibilities. The first one describes a policy in which the Running-Total is accumulated only for accounts of less than $1:

===

The Ambiguous Policy Resolved

Access the Customer-Account-Record.
If Account-Status is Delinquent,
    Write Account-Number to Officer-Action-Report.
Otherwise,
    If Account-Balance is less than $1,
        Set Account-Status to Pending.
        Accumulate Account-Balance into Running-Total.

===

That is quite different from a policy of accumulating a total to include the balance of all but delinquent accounts:

===

The Ambiguous Policy Resolved Differently

Access the Customer-Account-Record.
If Account-Status is Delinquent,
    Write Account-Number to Officer-Action-Report.
Otherwise,
    If Account-Balance is less than $1,
        Set Account-Status to Pending.
    Accumulate Account-Balance into Running-Total.

===

Finally, there is the third possibility, in which the running total is made up of all account balances:

===

The Ambiguous Policy Resolved a Third Way

Access the Customer-Account-Record.
If Account-Status is Delinquent,
    Write Account-Number to Officer-Action-Report.
Otherwise,
    If Account-Balance is less than $1,
        Set Account-Status to Pending.
Accumulate Account-Balance into Running-Total.

===

There are three possible Structured English conventions to delineate block boundaries:

1.   Mark the blocks explicitly by drawing boxes around them (as
in the example of Fig. 77) or by using some specific catch-
phrase.

2.   Use a military numbering or outline format; resetting the paragraph number shows the end of the block.

3.   Use an indentation convention as in the preceding examples.

The second and third of these are the more frequently used. The following is
an example in which an outline-like format helps to resolve the boundaries:

==

A Variation of the Ambiguous Policy Resolved the Third Way

1. Access the Customer-Account-Record.
2. If Account-Status is Delinquent,
       2.1 Write Account-Number to Officer-Action-Report.
   Otherwise,
       2.2 If Account-Balance is less than $1,
          2.2.1 Set Account-Status to Pending.
3. Accumulate Account-Balance into Running-Total.

==

Note that the "Otherwise" did not get a new number. Block number 2 is the
entire Decision construct, both the If part and the Otherwise part.

Whatever you use for block delineation (indentation, numbers, boxes)
will serve to show subordination as well. In the example above, the subordinate policies 2.1 and 2.2 are nested inside the Decision construct of block 2.
Both the indentation and numbering scheme used call attention to this nesting.

16.6 The balance sheet on Structured English

I sometimes feel that analysts are all extremists on the subject of Structured English. There are those who can barely restrain their enthusiasm for
this wonder which promises to add some degree of formalism and rigor to the
specification process. And then there are those who are unable to consider the
topic without audible groans of distress. Both are extreme. As the soul of un-
biased reason and moderation, I have taken it upon myself to set out the balance sheet on Structured English:

The Advantages of Structured English

1.   *It survives the life of the project.* During the analysis phase, it
can be used to describe policy; during module design, it can be
used to describe logic. It may also prove useful for odds and

ends of procedure description. Project members who become facile with Structured English tend to use it for everything.

2.   *It can be kept in automated format.* You can use text editors to keep it up-to-date. It can eventually be included in the listing of some program that implements the policy described. You may even get the maintenance programmer to modify it (since it is staring him in the face) when he makes a required change to the code.

3.   *It can be made concise, precise, and readable.* Structured English is probably as close as we will ever get to a formal specification language.

4.   *It can be tailored to suit the user.* Since our users range in EDP sophistication from total novices to computer jocks, it is essential that our specification tools have this flexibility.

5.   *It can be coordinated to the Data Dictionary and the Data Flow Diagrams to check consistency.* This is what I have been calling a lexical check.

6.   *It can be written quickly and naturally.* Once you have built some facility in Structured English composition, it is as simple as any form of writing. Since it is abbreviated and free from considerations of personal style, you will typically spend less time with it than you would with narrative description. This is important since we need to spend as little time as possible on each iteration so that we can go through more iterations.

There are some disadvantages, too. I believe they can all be overcome, but just for the record, here they are:

Disadvantages of Structured English

1.   *It takes some time to build Structured English skills.* While you are new at it, you may find yourself struggling to distinguish between policy and procedure, struggling to reduce redundancy, struggling to live within the syntax and vocabulary limits, or just plain struggling.

2.   *It seems to be more formal than it is.* Structured English is not a formal specification language. It is not rigorous. What it is, is terse. Descriptions written in Structured English tend to be simple and easy to read. But there is no guarantee that just because they can be expressed in Structured English, they are right.

3.   *It can scare off your user.* If you show him a variant that seems too foreign to him, he is liable to become a reactionary. He may refuse to look, ever again, at anything but Old English script.

16.7 Gaining user acceptance

I once had a student remark that Structured English seemed like a great tool for analysts to use among themselves and in communicating with the implementors, but not something to be shown to the user. I came to the melancholy conclusion that I had wasted that student's time. In my opinion, something you can't show to the user is *totally worthless* as an analysis tool. We have never had an insurmountable communication gap among analysts or among analysts and implementors. But there is a veritable communication chasm between the analyst and the user. It is this chasm that Structured Analysis is trying to bridge. In order to be of any use to us during the analysis phase, Structured English has got to be made acceptable to the user.

I have made the point above that part of the task of gaining user acceptance involves tailoring your Structured English conventions to his tastes and abilities. But presentation of the idea is equally important. A long pompous memo touting the glories of Structured English and ranting on about orthogonality, limited syntax, the Bohm and Jacopini constructs, and the like, is not the way to broach the topic.

Remember, you are not trying to tell the user that specifications will be written in Structured English. You are trying to convince him that there is no better way, and that he would be crazy to settle for anything less. I offer these points to help you sell this idea:

1.   Pick your formats first. Don't show him several and ask him to choose. The more abbreviated ones might frighten him, and he might then sense that the others are just disguised versions of the same thing. You have to guess — and guess right! — what will be acceptable to him.

2.   Let him see you working with Structured English before he realizes that it will be used in his specification. Let him review your notes. Solicit his observations about various policies written in Structured English. By the time you discuss the makeup of the specification, he must already be sold on your descriptive techniques.

3.   Whatever you do, *don't use any highfalutin jargon.* Don't even use the term "Structured English." If the user asks you what you call your descriptive language, tell him it's called English. Tell him you indent funny. Tell him you have a dull writing style. Tell him you hate adjectives.

4.   Show it to him a little at a time. Be present when he looks at it. Use a walkthrough to help him along.

5.   Work with one user at a time. Don't let them gang up on you.

Finally, if you find that you have to retreat from using Structured English because your user simply refuses to accept it, don't retreat further than you must. If you can't use Structured English, structure your use of English. Use Data Dictionary names wherever you can. Avoid compound sentence structures. Write so as to interrupt the reader's serial thought process as little as possible. Cross out every adverb and adjective that can be deleted without changing the meaning. Use indentation and numbering freely to enhance readability.

I said at the beginning of this section that if you partitioned enough, you couldn't get into too much trouble writing your mini-specs in narrative text. So if you are reduced to that, don't stop partitioning until your primitives are truly tiny.

If you judge that you will never be able to slip Structured English past your user and you must rely on text description, then I have some added homework for you. There is one very short and utterly delightful handbook on writing cited in the Bibliography (Flesch, *The Art of Plain Talk*). Buy a copy and read the first ten pages — that should be enough to get you hooked. He, too, has tried to settle on a clean and minimized subset of the English language for clear exposition. His subset is called Plain Talk. It is not too different from Structured English.

17 ALTERNATIVES FOR PROCESS SPECIFICATION

Certain kinds of policy simply cry out to be described using a Decision Table. When applied — and applied properly — to such a situation, a Decision Table can be unmatched for clarity and precision. In addition to being a descriptive tool, a Decision Table can help you to think out a policy in the making, to evaluate it for completeness and consistency. So if you find yourself with the task of writing a mini-spec for a natural Decision Table candidate, almost certainly no other means of describing it will be as useful.

Although the Decision Table is an important Structured Analysis tool, I will not discuss Decision Table techniques and their use in this chapter. If you have not had some previous experience with Decision Tables, I recommend Tom Gildersleeve's excellent book which is cited in the Bibliography.

This chapter will concentrate on how to recognize primitives that ought to be described by Decision Table, how to take advantage of the feedback possibilities that Decision Tables give you, and how to show them to your user. In addition, we will consider a slightly different way of approaching the topic in order to avoid some confusion. And, at the end, we'll discuss conventions for integrating Decision Tables with Structured English for policies that are only partially describable by Decision Tables.

17.1 When to use a Decision Table

Suppose you query your user about his policy for charging charter flight customers for certain inflight services, and he tells you something like this:

> If the flight is more than half-full and costs more than $350 per seat, we serve free cocktails unless it is a domestic flight. We charge for cocktails on all domestic flights . . . that is, for all the ones where we serve cocktails. (Cocktails are only served on flights that are more than half-full.)

Even if your experience as an analyst only goes back to 9 A.M. this morning, such a policy description must surely make you uncomfortable. You sense that it is likely to be incomplete or even self-contradictory. The text policy description makes it difficult to understand the whole policy, or even to put your finger on its failings.

Expressing the policy in the form of a Decision Table resolves all of these problems:

	RULES							
CONDITIONS	1	2	3	4	5	6	7	8
1. Domestic	Y	N	Y	N	Y	N	Y	N
2. Over half-full	Y	Y	N	N	Y	Y	N	N
3. Over $350	Y	Y	Y	Y	N	N	N	N
ACTIONS								
1. Cocktails served	Y	Y	N	?	Y	?	N	?
2. Free	N	Y			N			

In particular, it points up that there are some glaring holes in the statement made by the user — some components of the whole policy that have not been explained at all.

The thing that made this policy a natural candidate for a Decision Table is that there were a confusing number of subpolicies, and selection of the applicable one was dependent upon not just one but a number of conditions. In general,

> Use a Decision Table when subpolicy selection depends upon combinations of conditions.

The advantages of using Decision Tables under these circumstances seem fairly clear: They help you document your understanding with the user in a way that almost defies misunderstanding; and they help rout out situations that have not been fully specified. In the latter case, the Decision Table prompts you to go back to the user with detailed questions to help you fill out the table. Or you make an assumption and show it to the user. If you have guessed correctly what his true policy is, he will sign off on it, probably never realizing that he hadn't made himself at all clear. If you have guessed incorrectly, he may abuse you a bit and refer you back to his "lucid and definitive opus" describing the policy. But eventually he will set you straight. When you're done, the Decision Table is complete and correct.

You would be hard pressed to find an analyst who would argue against the use of a Decision Table when called for. But there are some legitimate problems tied to their use:

1.   It is sometimes hard to know where to begin with a Decision Table formulation.

2.   Everybody *knows* how to build the Decision Table matrix (it
       has something to do with binary numbers or something), only
       nobody can remember exactly when the chips are down.

Both of these problems stem from the method used to teach people about Decision Tables in the past. In the following sections, I hope to give you a different
view of the subject so that you will never again have to remember arbitrary
rules to work with Decision Tables.

A third problem has to do not with Decision Table technology, but with
the more sociological concern of getting our poor beleaguered user to work with
yet another new tool:

3.   Users are sometimes unfamiliar with Decision Tables. They
       may be apprehensive about their use.

17.2 Getting started

A Decision Table is a tool to distinguish among and specify a set of n subpolicies, only one of which applies in any given situation. If you knew the
value of n for a policy, you would have a headstart in building the table. At
least you would know its shape. The value of n establishes the width of the
table, the number of subpolicies or rules.

How do you figure out from your user interview how many rules there
are? Go back to the user's statement of policy on cocktails served in flight, and
see if you could have predicted from that alone how many rules there would
have to be in the Decision Table.

If you concluded that the number of rules must always be some power of
two, you missed something. Consider this situation:

The subsidy to the cub scout pack is based on number of scouts,
rank, and length of membership. The subsidy for each scout is $25
for first-year members, $35 for second-year members, and $50 for
scouts who have been members longer than two years. In addition
to this base, each scout gets an extra subsidy of $10 if he has at-
tained the rank of Wolf, $15 if he has attained the rank of Bear, and
$20 if he has attained the rank of Lion — unless he does it in his
first year, in which case he receives $70.

What is the size of a Decision Table to describe the per-scout subsidy policy?

In order to approach this in a methodical fashion, we need to concentrate
on the conditions that affect policy. A given subsidy depends upon the values
of a set of decision variables. What are the decision variables for this policy,
and what can their values be? Just to make it simple, let's give a name to each
decision variable and to each possible value.

DECISION VARIABLE	ASSOCIATED VALUES
Rank	None (N)
	Wolf (W)
	Bear (B)
	Lion (L)
Membership	First Year (F)
	Second Year (S)
	Third or More (T)

There are two variables, and they can take on four and three values, respectively. So, how many rules shall we expect our Decision Table to require?

It is well to approach this problem by decomposition. Suppose that there were not two variables that applied, but only one. Suppose that only rank mattered. Rank takes on four possible values. How many rules will be needed in our Decision Table to describe this reduced (one variable) policy? The answer is four. Now how much does the addition of the second variable (membership) complicate the situation? If there were only two possible values of membership, then the addition of that variable to our Decision Table would double its size. We would then have eight rules. But membership takes on three possible values: F, S, and T. Instead of doubling the size of our table, it will triple it. The final table for the scout subsidy policy will have twelve rules:

RULES

CONDITIONS	1	2	3	4	5	6	7	8	9	10	11	12
Rank	N	W	B	L	N	W	B	L	N	W	B	L
Membership	F	F	F	F	S	S	S	S	T	T	T	T

ACTIONS												
Subsidy	25	35	40	95	35	45	50	55	50	60	65	70

Go back and see if you can deduce the general principle that determines how many rules a Decision Table will have as a function of the number of variables and values. Try out your conclusion on a policy which allocates sales commission percentage based on Volume, Prepayment, and Salary:

DECISION VARIABLE	ASSOCIATED VALUES
Volume	Greater than $10,000 (G)
	Less than or equal (LE)
Prepayment	More than half prepaid (M)
	Not more (N)
Salary	Under $12,000 (U)
	$12,000-24,000 (M)
	Greater than $24,000 (G)

The answer is that a Decision Table of twelve rules will be required to describe the commission policy. The general principle is:

> The number of rules required to complete a Decision Table is equal to the product of the numbers of values for all decision variables.

In this case there are three decision variables that take on 2, 2, and 3 values, respectively. The number of rules is therefore $2\times2\times3=12$.

17.3 Deriving the condition matrix

Since the number of rules depends only on the decision variables and their related values (not the resultant subpolicies), it follows that you can build your Decision Table's condition matrix entirely and then go back and fill in the actions. Separating the building of the matrix from determining the subpolicy has got to simplify the process.

When you write the condition matrix, you define each of the rules as combinations of settings of the various variables. You must be systematic in going about this or you will never get all the rules. In the above sales commission policy, for example, you might end up struggling to find the twelfth, or worse yet, you might end up with thirteen. If each of the variables can take on only two possible values, your system of laying out the matrix is very similar to counting with binary numbers. That may or may not make some sense to you. In any event, it is not very relevant here since at least one of our decision variables is "trinary."

If I gave you a procedure to lay out any condition matrix, regardless of number of variables and values per variable, you would probably forget it. It's the same procedure you have forgotten before. I would be a hypocrite to "teach" it to you since I would have to look it up myself. Instead, I offer an easy way to derive the procedure whenever you need it. All that is required is that you consider the variables one at a time.

Let's try this approach, using the sales commission policy as an example: First, assume that only one variable, Volume, matters in setting commission rate. How big a Decision Table would we have then? (Two rules) Draw that condition matrix for that Decision Table:

RULES

CONDITIONS	1	2
Volume	G	L

That would be the matrix if there were only one variable that mattered. But there is at least one other; there is the number of prepaid sales, for instance. That variable, Prepayment, takes on two possible values, so what will adding it to our matrix do? (It will double the matrix.) Double your current matrix by copying it over into rules 3 and 4. Fill in the two different values of the added variable for the two different halves:

RULES

CONDITIONS	1	2	3	4
Volume	G	L	G	L
Prepayment	M	M	N	N

That would be our entire matrix if only those two decision variables mattered. But there is yet another, Salary range. It takes on three values: U, M, and G. What is the addition of this last variable going to do to the size of our matrix? (Triple it.) Draw the tripled matrix by copying rules 1 through 4 over into 5 through 8 and then again into 9 through 12. Now go back and fill in the three values of the added variable for the three thirds of the matrix:

RULES

CONDITIONS	1	2	3	4	5	6	7	8	9	10	11	12
Volume	G	L	G	L	G	L	G	L	G	L	G	L
Prepayment	M	M	N	N	M	M	N	N	M	M	N	N
Salary	U	U	U	U	M	M	M	M	G	G	G	G

That is the complete matrix for the problem as defined. All that remains is to fill in the ACTIONS section, specifying the commission rate that applies for each rule.

17.4 Combining Decision Tables and Structured English

You may come across a policy that could use a Decision Table for part of its description, but that requires Structured English for the rest. (Usually this kind of complication is a sign that you didn't partition enough.) I see no reason why the two techniques couldn't be combined. You might have the Structured English inside the Decision Table, or vice versa. Here is one example:

	RULES			
CONDITIONS	**1**	**2**	**3**	**4**
Volume	G	L	G	L
Prepayment	M	M	N	N
ACTIONS				
Set Commission-Rate to	4%	5%	6%	8%
Set Commission-Amount to Product of Commission-Rate and Volume. If Salary greater than $12,000,				
Then increase Commission-Amount by	$100	$150	$150	$200

17.5 Selling Decision Tables to the user

A Decision Table might be useful to help you work out a complicated piece of logic, even if you decided that you couldn't show it directly to your user. But frankly, most of the advantage to be gained from the table would be lost. Remember, the logic you are working with is logic that pertains to the user's policy. It is not the analyst but the user who has the greater need for insight about the complicated policy, its effects and ramifications.

Communication across the user-analyst gap can be greatly facilitated by the use of Decision Tables. If the user is not familiar with the idea, you have to act as his "toolsmith" and help him along. If you manage to teach him some Decision Table skills and convince him of their value, you will both be ahead. At the very least, the user should be taught to read Decision Tables and helped to feel comfortable with them.

The same principles apply to your introduction of this new concept (if it is new to him) as have applied before:

- Use no jargon. (It need not be called a Decision Table; call it a "table that describes a particular policy" or a "situation.")

- Let him see your use of it informally before he ever suspects that it is to become part of the specification.

- Guide him through it. Work with him on it. Make him think it is a "working tool" (it is), rather than a documentation technique.

You can lose the battle before it has even begun if you go about it wrong. Fostering the idea that a Decision Table is your own private tool may make the user reluctant to look at it. When you present him with it, he may just give it a perfunctory glance and say something like, "Yes, well, that's very fine. If you have done it right, and properly included everything I wrote in my very clear memo on the subject, then it's right. And if you haven't, it isn't. I assume you've done it right." That is not sufficient as a user reaction. He has just signed a blank check for future hedging. In fact, the Decision Table has not communicated anything to him, because he hasn't read it. It is the very worst kind of user-analyst non-communication, because each retains his own understanding of what the policy is, together with the self-righteous feeling that the other guy had his chance to understand and it's his own damn fault if he still hasn't got it right.

You have to convince the user that the Decision Table *is* the policy. His "very clear memo on the subject" was only an input to the process and has no future relevance. If he doesn't read and understand the table, he can't be sure of what he's getting. Particularly if you have incorporated some assumptions (because his very clear memo wasn't so very clear), the user is obliged to verify each and every rule.

17.6 Decision Trees

A Decision Tree is a graphic representation of a Decision Table, nothing more, nothing less. It is useful for exactly the same kind of policy as Decision Tables are. I advocate the use of a Decision Tree only for a user who simply cannot be persuaded to look at a Decision Table. The Decision Tree is sometimes easier to swallow for such a user because it looks more familiar. (It is, after all, more or less a Family Tree, something that the user knows every bit as much about as any of those beard-and-sneaker boys in the systems department.)

Because of its familiar appearance and graphic presentation, a Decision Tree functions as a self-teaching tool. That being the case, I won't teach you about it, and you won't have to teach your user. A few examples ought to suffice. Presented in Fig. 78 is a Decision Tree version of the cub scout subsidy policy you looked at earlier in this chapter. And in Fig. 79, I have expressed the Sales Commission policy with a Decision Tree.

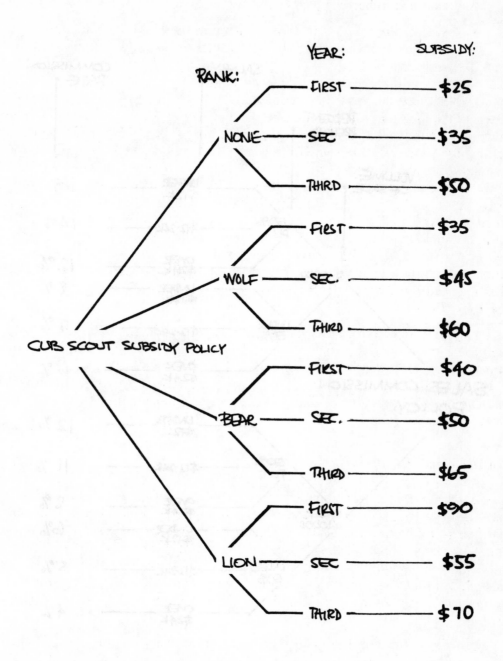

Figure 78

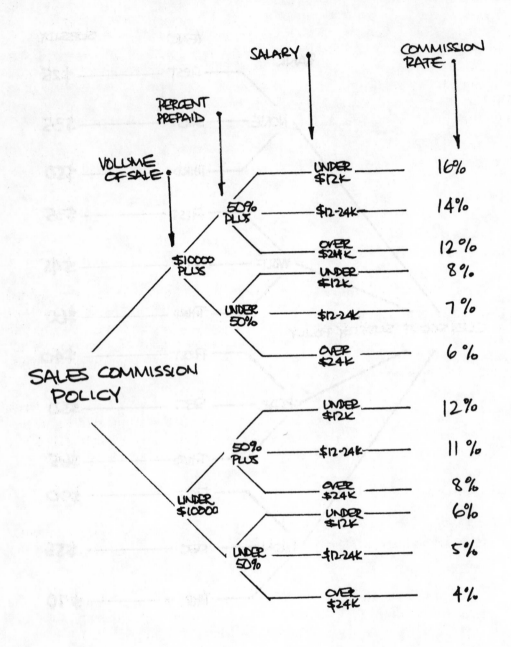

Figure 79

A Decision Tree is very like what has come to be called a Warnier Diagram, after Jean-Dominique Warnier. Warnier is a French system scientist who has consistently been ahead of his time. He was an independent inventor of most of the concepts of Structured Programming and some of the most important ideas of data-driven design. In his written work (see Bibliography), he has proposed the use of Warnier Diagrams for much more than just mini-spec writing. In particular, he finds them applicable to Data Dictionary, to program and module design, and to a host of other things. The 1977 translation of his 1961 book, *Logical Construction of Programs,* created a great stir among analysts and implementors, even though it was already more than 15 years out of date.

Recently, an American system scientist, Kenneth Orr, has out-Warniered Warnier in his use of the diagrams. For a very readable description of the extended use of Warnier Diagrams for analysis and design, I refer you to Orr's new book (also cited in the Bibliography).

17.7 A procedural note

Some analysts defer specification of the functional primitives until implementation time. They reason that, having isolated a given process and defined its interfaces to the rest of the world, they take no substantial risk by putting off a formal description of its governing policy. The advantage of deferring specification is that it can be performed as part of the task of preparing an implementation of the policy, thus saving some potential communication problems and some writing effort.

This is a radical approach; I am not trying to sell it to you, only to indicate that it is done. There is some risk, specifically that your incomplete understanding of underlying policy allocated to a particular process might cause you to miss an interface. On the other hand, the savings in time are attractive.

17.8 Questions and answers

Here we are again at the part where I ask myself questions frequently asked by my students, adroitly choosing the questions to which I know the answers:

> *Have you had actual experience with users who were not at all sophisticated about EDP, but who nonetheless learned to work effectively with Structured English?* Yes. The idea of formalistic notation is not totally foreign to anyone. A typical user might already be conversant with chess and bridge notations, special language subsets for auctions, for air traffic control, for CB communication, and the like. Chances are, the system you are specifying for the user will interface to him through some sort of formal linguistic facility. So Structured English fits right in. I believe that a failure to sell the concept of Structured English to a user is just that, a failure.

Much of the Structured English you showed in your examples was practically at a one-to-one level with code. Doesn't that seem wrong? I don't think it has much relevance. If five lines of Structured English end up causing you to write five lines of COBOL, it only means that COBOL turned out to be a pretty good language to use to code that particular module; i.e., the code was fairly conceptual. It certainly does not imply that you should have been writing a much more abstract level of Structured English, so that you ended up with a 10:1 or 30:1 relationship. You write Structured English to describe user policy, so it is as abstract or as detailed as the policy itself.

Isn't your Structured English description going to preempt part of the function of the coder? To the extent that form follows function, he may write his code to respect the pattern of the Structured English description of the policy to be implemented. I see no problem with this. On the contrary, it will make the code more readily comprehensible to the maintenance staff.

If the coder follows your punctuation in a language like COBOL, he can get in trouble. That is true. It is awkward and unfortunate. (But then, if you're used to COBOL, you're used to awkward and unfortunate.) I haven't got a good way to resolve this problem. You can't ask the analyst to respect COBOL punctuation standards. You could write Structured English with no punctuation, but that costs you something in clarity.

You didn't say anything about reduced Decision Tables. No. Or about multi-page Decision Tables. I suggest you not use them. The more complicated features of Decision Table technology apply mostly to the case of primitives that are too big. I would hope you would deal with this problem by going back to the DFD's and partitioning down another level.

Isn't there a compiler that generates code directly from Decision Tables? Yes, it's called DTAB67. The 67 stands for the date of development; it also stands approximately for the date at which its popularity peaked. Since only a small portion of the logic in most systems lends itself well to description by Decision Table, there is not much advantage to direct compilation. There are, however, some handy cookbook procedures for coding Decision Tables. (See Ed Yourdon's *Techniques of Program Structure and Design,* cited in the Bibliography.)

Would you ever want to use both Decision Tables and Decision Trees in a specification? I can't see why. If your user is willing to live with Decision Tables, you might as well use them throughout.

PART 5

SYSTEM MODELING

18 USE OF SYSTEM MODELS

Up to this point, we have looked at all of the tools of Structured Analysis and their use. What we have not dealt with, except indirectly, is the "how to" of Structured Analysis, i.e., how to put the various tools to work in completing the analysis phase and, in particular, how to apply them in developing the Structured Specification. This is where we go back and fill in some of those holes.

Fig. 80 is a Data Flow Diagram that you have seen before. It served as the original definition of Structured Analysis at the very beginning of this book. The first process shown in the figure (Bubble 2.1: Study current environment)

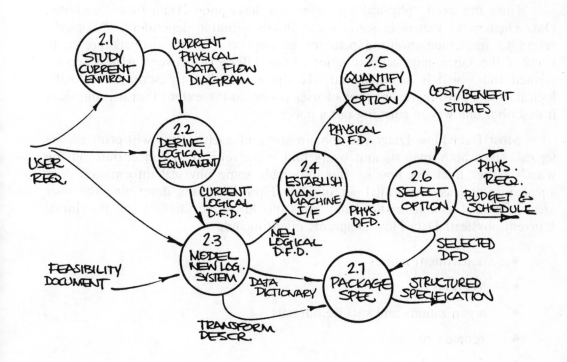

Figure 80

involves drawing a set of leveled Data Flow Diagrams to document the current way of doing business in the user area. (In most cases, there is at least a partial Data Dictionary compiled as well as part of this process.) Once the current physical description is completed, the rest of the analysis phase can be thought of as a series of transformations:

- transformation of the current physical description to its logical equivalent (Bubble 2.2)

- transformation of the current logical to the new logical (Bubble 2.3)

- transformation of the new logical to the new physical (Bubble 2.4)

The majority of analysis phase manpower is spent on these transformations. Each one advances us further toward our goal of producing a Structured Specification. When all three transformations are done, only a packaging step is necessary in order to complete the specification.

18.1 Logical and physical DFD characteristics

I use the word "physical" to refer to a description (Data Flow Diagrams, Data Dictionary) that is in some sense implementation-dependent. "Logical" refers to implementation-independent description. Some analysts prefer to think of the logical-physical distinction as the difference between *what* is accomplished and *how* it is accomplished. To the extent that a description is totally logical, it is a pure representation of user policy; to the extent that it is physical, it describes one way of carrying out a policy.

Most Data Flow Diagrams contain some of each. They will primarily be logical, since both analysts and users find it convenient to concentrate on the what's rather than the how's. But invariably some physical information does sneak into at least the initial set of Data Flow Diagrams describing the user area. Physical characteristics that you should expect to see in the initial (current physical) Data Flow Diagrams might include

- department names

- locations

- organizations and service bureaus

- people's names

- files and data stores

- procedure

- devices

- existing automated and electromechanical systems

No one ever sets out to include great numbers of physical characteristics in a Data Flow Diagram, but they are often useful.

1.   They may help us get started in our partitioning. If a corporate organization has been carefully thought out, then adhering to organizational lines in the initial Data Flow Diagramming may yield a very serviceable and easily obtained result. After all, the organizers of a corporation have the same partitioning goal that we analysts have: to minimize interfaces.

2.   Physical characteristics in the Data Flow Diagram may help elicit a meaningful concurrence from the user. He is likely to require at least some of these physical "benchmarks" in order to relate a Data Flow Diagram back to his operation.

3.   It may be easier to describe something physically than to ferret out the underlying policy.

All of these are good reasons to allow some physical elements in our Data Flow Diagrams. It often happens that additional physical considerations intrude simply because they are not recognized as being physical — there is no well-defined binary distinction between logical and physical; there is rather a continuum of characteristics from the most purely logical to the most obviously physical.

Since physical attributes are forever cropping up in Data Flow Diagrams, and since they are often quite useful, Structured Analysis adopts this policy toward them: *Physical attributes are allowed in the initial subphase of analysis and then systematically eliminated later.* The product of that first subphase is termed the current physical model. It is reviewed with the user to elicit his agreement that it is a proper representation of his current environment. Then, after we have taken full advantage of the physical attributes, we remove them (this is the current-physical to current-logical transformation) to prepare an equivalent logical description.

18.2 Charter for Change

The first two subphases of analysis (Bubbles 2.1 and 2.2 of Fig. 80) take their input from the user alone and are completely concerned with the current environment. As an analyst goes about this stage of his work, he is quite aware that there is some change pending: some set of procedures is to be automated, or centralized, or decentralized, or new functions are to be added, or something of the sort. This new set of rules to be applied to the future environment is what I refer to as the "Charter for Change."

Note in Fig. 80 that the Charter for Change is introduced as part of the Feasibility Document (output of the survey phase). It has no effect on the early part of analysis. It is the Charter for Change that distinguishes the new logical system from the current logical one that it will replace. The Charter for Change is a major input to the second transformation, the current logical to new logical (Bubble 2.3). It is during this transformation that the logical model of the new system is produced. The model consists of an integrated set of Data Flow Diagrams, Data Dictionary, and mini-specifications.

18.3 Deriving the Target Document

The new logical Data Flow Diagrams together with supporting documents constitute the most logical expression of the new system. They state what has to be done, but little or nothing about how. The whole rest of the project (the rest of analysis plus all of design and implementation) is concerned with the further "physicalization" of this system.

The system described in our logical model will eventually be implemented with some automated and some human procedures. But the model does not distinguish between automated and human procedure — it states the functions being performed but does not allocate them to the inside or outside of the computer system to be built.

The new logical model is a valuable interim analysis product, but it is clearly *too logical* to be our Target Document. In order to come up with the Target Document (in the form that I have referred to as a Structured Specification), some physical information must be incorporated. At the very least, we must establish where the man-machine boundary shall be: which logical procedures have to be automated and which should remain manual.

The third transformation, the new logical to new physical transformation, involves adding the minimum physical information to establish true project targets. This might include

- the man-machine boundary
- operational characteristics
- performance targets

Note that the final result is still not very physical. It is important to give the implementation team maximum leeway by excluding anything that is not a true project target from the Structured Specification. If there is more than one acceptable way to satisfy a target, we should endeavor to leave the choice to the designers and implementors. Two possibilities may seem equivalent to the analyst; but the coder, from his very different perspective, might know that one is substantially easier to implement than the other.

In the next chapter, we'll look at the three transformations in detail.

19 BUILDING A LOGICAL MODEL OF THE CURRENT SYSTEM

The first transformation involves removing all the physical information in the current model. An example will get us started: The Data Flow Diagram of Fig. 81 presents a simple freight-forwarding operation. The figure is full of physical considerations, but it may nonetheless be a useful working document since it is easy to derive and easy for the user to relate to. As analysis proceeds, however, the physical considerations become a burden. They become less and less necessary to facilitate communication with the user, and they cause some problems of their own. They may introduce arbitrary constraints, requirements that ought not to be imposed on the rest of the project. Use of physical terms also limits the readership of the documents to those who are familiar with the details. (In the example of Fig. 81, not knowing what a Form-448 is makes it hard to understand what is going on.) Having taken advantage of the physical nature of the early working models, it is now time to clean them up and make them logical.

Any physical aspect of a current model is likely to fall into one of four categories: It is likely to be political, procedural, historical, or tool-related. Fig. 81 has some physical characteristics of each variety.

The partitioning of function between the front office and the treasurer's department, for instance, is a good example of a political consideration. The data flow called President's-Weekly-Report is also political, in that it refers to a particular position in the organization.

Any of the data flows that are made up of forms or explicit document names are probably procedural in character. The data flow called "Yellow-448-Copy" is an example of this. The required information from that form might only be a dollar value or tax amount or some such thing, but for the sake of convenience a complete copy is routed. Hence, the data flow is physical; its composition is affected by a particular procedural approach to carrying out policy. The fact that the Shipping-Manifest is routed through the Treasurer's Department and then to the file, rather than vice versa, is another procedural detail. It happens to be done that way, but it doesn't *have* to be done that way.

The historical type of physical detail is often harder to spot. In Fig. 81, you might suspect that there is some historical reason why the Work-In-Progress-File consists of odds and ends rather than a set of separate files, each

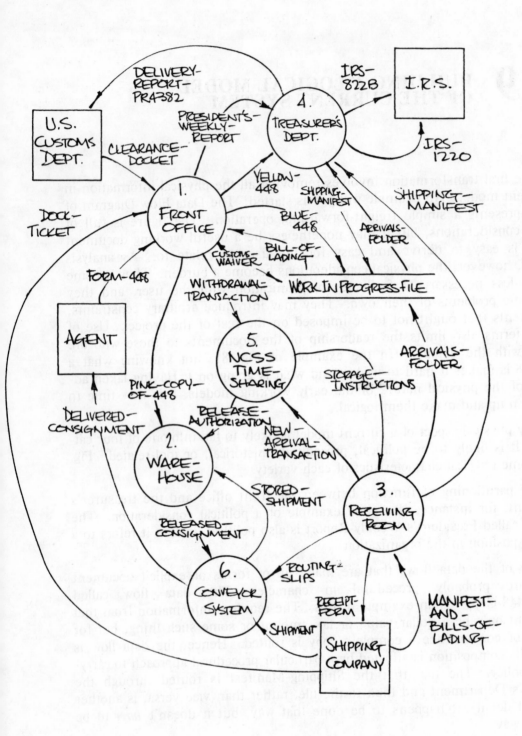

Figure 81

with its own allocated use. When you look inside the file, you might also find that there are data elements that are no longer required, or data groupings that seem arbitrary and capricious, both indications of some historical relationship.

The conveyor system and time-sharing system bubbles are clearly tool-related. They are derived from particular hardware or hardware-software combinations that are used to carry out policy.

As you can see, any of the DFD elements (process, data flow, or file) can take on a physical nature. All will have to be revised as part of the transformation. The derivation of logical equivalents proceeds in this fashion:

1.   Build an expanded DFD to remove the highest level(s) of physical characteristics.

2.   Use decomposition and normalization to describe the file structure in its most logical form.

3.   Work bottom-up to eliminate procedural and historical characteristics by minimizing data flow.

A section below is dedicated to each of these procedures.

Before we go into the details, I want you to know that the derivation of logical equivalents is never a total and unqualified success. In some cases, it is impossible to divorce the policy described by your model from an implementation of the policy. In general, though, it will be possible to make a considerable improvement in the model by repeated application of the three "logicalization" procedures.

19.1 Use of expanded Data Flow Diagrams

The greatest concentration of physical characteristics is in the top few levels of the DFD set. In particular, the political and tool-related physical aspects are present only at or near the top. But bubbles at the upper levels are simply representations of the networks at the next level down. You could eliminate any one of the bubbles at level n by replacing it with the n+1 level network it represents. To the extent that the replaced bubble was physical, such a replacement ought to improve the logical character of the diagram.

Consider the freight-forwarding example of Fig. 81. Bubble 5, the "NCSS Time-Sharing System," introduces a strongly physical consideration into the figure. But the associated level-two diagram (shown in Fig. 82) is not nearly so physical. It is concerned with the functions allocated to the time-sharing system. Replacing Bubble 5 on the parent with the lower-level description of the same subject matter makes our resultant diagram more logical. In this case, replacing Bubble 5 by the network it represents is a replacement of a physical component by its logical equivalent.

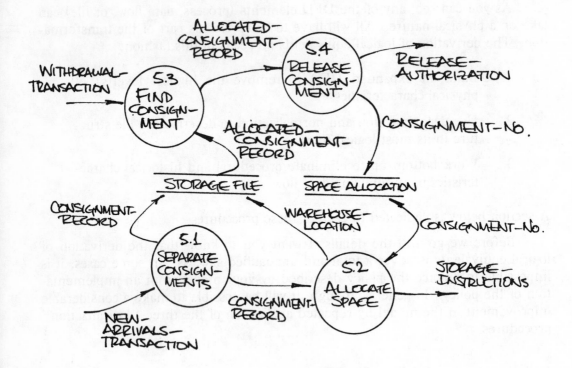

Figure 82

Fig. 83 is the partially expanded DFD resulting from the replacement of Bubble 5. Now, in a similar fashion, you could replace Bubble 1 by its corresponding level-two network. Again, that ought to be an improvement. Since all the top-level bubbles are highly physical (at least in this case), you could profit from replacing all of them by their level-two networks. The final diagram would be rather cluttered, but it would be much more logical.

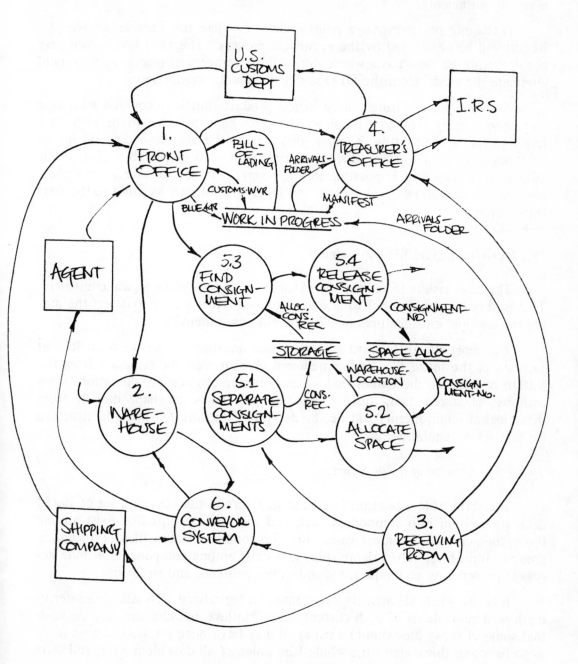

Figure 83

The same idea might be used again, replacing level-two bubbles by level-three networks to eliminate further physical considerations. If you go far enough, you will construct the bottom-level diagram, the football field of paper that is a network of functional primitives. But normally you will not need to go very far down in this process before the diagram is rid of all the political and tool-related bubbles. Replacement of level-one and selected level-two bubbles is usually sufficient.

It is quite possible that a political data flow like the President's-Weekly-Report will be eliminated by the expansion process. The next level down may not deal with the report as a whole, but with its component pieces; so you could eliminate the report from the DFD and just show the components.

Now you have a much more logical product, but it is covered with tape and has so many crossed data flows that you have to trace them with your finger to follow their paths. Clearly, this expanded Data Flow Diagram has to be thought of as a working document only, not as a deliverable. Since it is difficult to update and impossible to photocopy, it ought to have a very short lifetime. In fact, it will — the expanded DFD is used only as input to the next transformation.

19.2 Deriving logical file equivalents

The next task is to replace the existing set of files by its logical equivalent. That is very easy to say, but what does it mean? How are we to detect the procedural and historical features of files and eliminate them?

The most straightforward answer to that question is that we must look at each *use* of the set of files and somehow derive an ideal file structure from the pattern of use. To the extent that we can remove physical considerations from each use of stored data, the file structure that takes its shape directly from these logical requirements will be, by definition, the most logical file structure for that set of requirements.

19.2.1 The Concept of a File Structure

The term "file structure," as I use it, refers to the complete set of stored data, its division into component sets, and the relationships that exist among those components. In most cases, the component sets are files or repeating groups within files. The relationships that exist among components include access keys, ordering techniques, redundancies, pointers, and so forth.

It is the *whole* file structure that must be logicalized. To talk about deriving logical equivalents of each current file is fruitless, because the very decision that some of these files should exist at all may have been physical. So we have to go back and think about the whole (the union of all data elements), and start from scratch to divide this "superfile" into its component pieces.

The superfile for our freight-forwarding example would be the union of all the data elements in any of the three files shown in the partially expanded DFD of Fig. 83. The Data Dictionary definitions for those files follow:

Work-In-Progress	=	*File of all info about received goods*
	=	{Ship-Name + Arrival-Folder + (Manifest)}
Arrival-Folder	=	{Bill-of-Lading +
		(Allocation-Ticket) +
		(Customs-Waiver) + (Form-448)}
Bill-of-Lading	=	Consignment-No. +
		Destination +
		Shipped-Weight +
		Agent-Name +
		Ship-Name +
		Consignment-Value +
		{Parcel-No. +
		Parcel-Description +
		Parcel-Value +
		Customs-Code}
Allocation-Ticket	=	Consignment-No. +
		Warehouse-Location +
		Received-Weight
Customs-Waiver	=	Consignment-No. +
		Authorization-Code
Form-448	=	Consignment-No. +
		Agent-Name +
		{Parcel-No. +
		Parcel-Description}
Manifest	=	Ship-Name + Arrival-Date +
		{Consignment-No. + Consignment-Value}
Storage-File	=	*File showing how consignments stored*
	=	{Consignment-No. +
		Number-of-Spaces-Allocated +
		{Warehouse-Location +
		{Parcel-No.} } }
Space-File	=	*File of space allocation in warehouse*
	=	{Warehouse-Location +
		Consignment-No.}

As you can see, the file structure of our current model consists of seventeen data elements grouped into three files with several internal repeating groups and virtually uncontrolled redundancy. It is from this physical file structure that we must derive the logical equivalent.

19.2.2 Decomposition of the File Structure

We begin the derivation of a logical file structure by considering each access to the file:

First Step: Compile a census of all references to stored data. For each reference, record data flow, direction, and purpose.

If you look back at Fig. 83, you will see that there are twelve stored data references. They are documented by a census of this form:

CENSUS OF PHYSICAL ACCESSES

ACCESS	PROCESS	READ/WRITE	DATA FLOW AND PURPOSE
1.	1.	Write	Store customs waiver upon receipt of clearance docket.
2.	1.	Read	Fetch bill of lading and customs waiver to verify release to agent.
3.	1.	Write	Store blue copy of Form-448.
4.	3.	Write	Store arrivals folder upon receipt of new shipment.
5.	4.	Write	Store shipping manifest along with arrivals folder.
6.	4.	Read	Fetch arrivals folder to prepare report PR4382.
7.	5.1	Write	Store consignment-no. and agent's name for newly received consignment.
8.	5.2	Read	Find unallocated space(s) for new consignment.
9.	5.2	Write	Mark space(s) used by consignment.
10.	5.2	Write	Mark consignment allocated to space(s).
11.	5.3	Read	Find location allocated to consignment.
12.	5.4	Write	Mark space(s) released upon removal.

Fig. 84 is an annotated copy of the previous figure; I have written in the access numbers to help you keep track.

Let's look at access 2, the fetching of a Bill-of-Lading, and, if it has arrived, the attached Customs-Waiver. I have not shown you the lower levels of Bubble 1 where this fetched data is put to use, but if you had access to the lower-level figures and to the associated mini-specs, you would quickly come to the conclusion that not all the data in the fetched data flow is necessary. The whole purpose of the access is to decide whether a given consignment ought to be released to the agent who has just presented a Form-448 requesting it. For such a determination, only the items Agent-Name (from the Bill-of-Lading) and Authorization-Code (from the Customs-Waiver) are required.

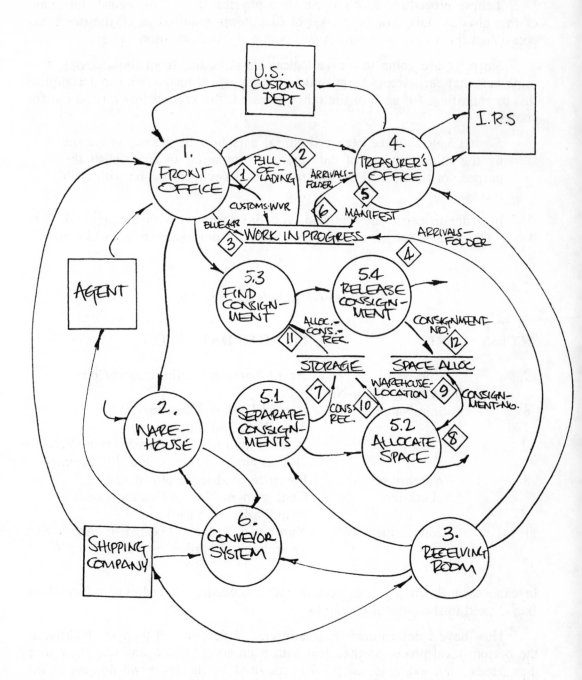

Figure 84

The physical data flow of access 2 is Bill-of-Lading + Customs-Waiver. Use of these procedural forms is clearly a physical trait. The logical equivalent of that physical data flow — the set of data items required to fill the need that occasioned the access — is only Agent-Name + Authorization-Code.

Since we are going to derive logical file structure from the accesses, it is important that each access be put into its most logical form. We can accomplish this by repeating, for each of the other accesses, the process just carried out for access 2.

Second Step: Replace each procedural data flow in the access census by the minimum set of data elements required to accomplish the purpose of the access. Associate required key information with each access.

In order to carry out this step, you have to start with the *incoming* data flows. The revised census of input data flows for our example looks like this:

CENSUS OF LOGICAL READS

ACCESS	KEY	LOGICAL DATA FLOW
2.	Ship-Name + Consignment-No.	Agent-Name + Authorization-Code
6.	Ship-Name	Arrival-Date + Company-Name + Total-Value-on-Ship + {Consignment-No. + Consignment-Value + {Parcel-No. + Parcel-Value + Customs-Code} }
8.	Warehouse-Location	Presence or absence of a stored Consignment-No. (indicating whether or not space is in use)
11.	Consignment-No.	Number-of-Spaces-Allocated + Agent-Name + {Warehouse-Location + {Parcel-No.} }

In each case, I simply have recorded the accessing or ordering information (keys) used in the current operation.

How have I determined these subsets? I determined them by looking at the bottom-level processes that deal with each access and examining their true data needs. For example, access 6 is required by the level-two process called "Produce Customs Alert Survey." A segment of the Data Flow Diagram containing this process is presented in Fig. 85.

Production of the summary involves using the list of ships that arrived during the week (information taken from the President's-Weekly-Summary) to access the proper arrival folders from the Work-In-Progress-File. The output data flow, Form-PR4382, is a report on the week's arrivals for use by the Customs Department. Its purpose is to convey to that authority data about values of received consignments, grouped by customs code and summarized by ship. The Data Dictionary definitions describing PR4382 are reproduced below:

PR4382   =   *Customs alert summary report*
                                  =   {Ship-Heading + {Detail-Line} }

Ship-Heading   =   Ship-Name + Arrival-Date +
                                       Company-Name + Total-Value-on-Ship

Detail-Line   =   Consignment-No. + Consignment-Value +
                                       Declared-Value-for-Customs-Code-"A"-Items +
                                       Declared-Value-for-Customs-Code-"B"-Items +
                                       Declared-Value-for-Customs-Code-"M"-Items +
                                       Declared-Value-for-Customs-Code-"R"-Items

With this information in hand, you can verify that the logical data flow specified for access 6 is indeed sufficient for generation of the report. The remaining information in the Arrival-Folder (the physical data flow that was accessed by the process) is extraneous.

Once the census of logical input data flows is complete, it is possible to derive the census of logical outputs. Again, each procedural data flow must be replaced by the data flow that is truly required. In order to determine what output data flow is required for each access, look back at the census of inputs. There is no use putting something into a file if it is never to be taken out.

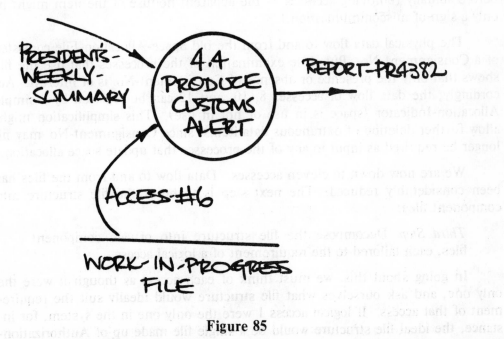

Figure 85

By this line of reasoning, the following census of logical outputs is derived for our example:

CENSUS OF LOGICAL WRITES

ACCESS	KEY	LOGICAL DATA FLOW
1.	Consignment-No.	Authorization-Code
3.		None
4.	Ship-Name	{Consignment-No. + Agent-Name + Total-Value-on-Ship + {Parcel-No. + Parcel-Value + Customs-Code} }
5.	Ship-Name	Arrival-Date + Total-Value-on-Ship + Company-Name
7.	Consignment-No.	Agent-Name
9.	Consignment-No.	Number-of-Spaces-Allocated + {Warehouse-Location + {Parcel-No.} }
10.	Warehouse-Location	Presence of Consignment-No.
12.	Warehouse-Location	Presence of Consignment-No.

Since the information on Form-448 is not used by anyone, access 3 is extraneous to our model and can be removed. (You might check with your user before actually removing accesses — the apparent nonuse of the item might be only a sign of miscommunication.)

The physical data flow to and from the old Space-Allocation-File consisted of a Consignment-No. But close examination of the processes that use the file shows that only the presence or absence of Consignment-No. is of interest. Accordingly, the data flow of accesses 8, 10, and 12 can be replaced by a simple Allocation-Indicator (space is in use or not in use). This simplification might allow further deletion of extraneous data items, since Consignment-No. may no longer be required as input to any of the processes that update space allocation.

We are now down to eleven accesses. Data flow to and from the files has been considerably reduced. The next step is to break the file structure into component files:

Third Step: Decompose the file structure into private component files, each tailored to the requirement of a logical access.

In going about this, we must think of each access as though it were the only one, and ask ourselves what file structure would ideally suit the requirement of that access. If logical access 1 were the only one in the system, for instance, the ideal file structure would be a single file made up of Authorization-Codes, accessed by Consignment-No. This is Private-Component-File (PCF) 1, and it exists only to serve the needs of logical access 1.

The complete set of private component files for our example is as follows:

PCF1 = {Consignment-No. + Authorization-Code}

PCF2 = {Ship-Name + Consignment-No. +
Agent-Name + Authorization-Code}

PCF4 = {Ship-Name +
{Consignment-No. + Agent-Name +
Consignment-Value +
{Parcel-No. + Parcel-Value +
Customs-Code} }

PCF5 = {Ship-Name + Arrival-Date +
Total-Value-on-Ship + Company-Name}

PCF6 = {Ship-Name + Arrival-Date +
Total-Value-on-Ship + Company-Name +
{Consignment-No. + Consignment-Value +
{Parcel-No. + Parcel-Value +
Customs-Code} } }

PCF7 = {Consignment-No. + Agent-Name}

PCF8 = {Warehouse-Location + Allocation-Indicator}

PCF9 = {Consignment-No. + Number-of-Spaces-Allocated +
{Warehouse-Location + {Parcel-No.} } }

PCF10 = {Warehouse-Location + Allocation-Indicator}

PCF11 = {Warehouse-Location + Allocation-Indicator}*DUPLICATE*

PCF12 = {Warehouse-Location + Allocation-Indicator}*DUPLICATE*

There is no PCF3 because access 3 was found to have no logical data flow.

It is something of a sublime fiction even to consider a structure made up of input-only and output-only files. The data are left on their own to move around among the files (perhaps by osmosis), so that a write to PCF1 magically causes the next read from PCF2 to have an updated value of Authorization-Code. Please bear with me for the moment on this. Use some Constructive Wishful Thinking to accept my assurance that the problem of interfile communication will sort itself out.

19.2.3 Normalization of Component Files

The PCF file structure is based on the arbitrary assumption that one file be established for each logical access in the census. If we allow the possibility of two or more files together filling the needs of a single logical access, we will be able to break some of the PCF's down further. Consider PCF6:

PCF6   =   {Ship-Name + Arrival-Date +
                       Total-Value-on-Ship + Company-Name +
                       {Consignment-No. + Consignment-Value +
                         {Parcel-No. + Parcel-Value + Customs-Code} } }

Private-Component-File 6 *happens* to have a number of subordinate records (one per consignment on the ship) included within each ship's record. But it doesn't *have* to be that way. We could, for instance, break it into two files, one recording the attributes of the ships, and the other keeping information about consignments on those ships:

PCF6′   =   {Ship-Name + Arrival-Date +
                       Total-Value-on-Ship + Company-Name}

PCF6″   =   {Ship-Name + Consignment-No. + Consignment-Value +
                       {Parcel-No. + Parcel-Value + Customs-Code} }

The two files PCF6′ and PCF6″ are together equivalent to the old PCF6. They were formed by removing the large internal repeating group from PCF6 and setting it up in a separate file. The residual file (PCF6 minus its internal repeating group) became PCF6′. The isolated repeating group became PCF6″. (Note that PCF6″ is not exactly the same as the internal repeating group: Its key has been modified to retain the correlation between ships and consignments on those ships. Without this key modification, the two new files would not be completely equivalent to PCF6.)

The process of removing internal repeating groups from complex files and setting them up separately is called *normalization*. Each time you normalize a file, you replace a multipurpose file by two or more files, which together can accomplish the same set of purposes. The new files are always more single-minded than the one they replace.

It frequently happens that the new equivalent files need to be normalized again. PCF6″, for instance, still has an internal repeating group. The normalization process needs to be applied one more time to spin off the information about parcels into their own file. With that second normalization, we arrive at this set of three files, which together are the equivalent of PCF6:

PCF6′   =   {Ship-Name + Arrival-Date +
                         Total-Value-on-Ship + Company-Name}

PCF6‴   =   {Ship-Name + Consignment-No. + Consignment-Value}

PCF6⁗   =   {Ship-Name + Consignment-No. + Parcel-No. +
                       Parcel-Value + Customs-Code}

The next step in our logical file derivation process requires that each of the PCF's be normalized:

Fourth Step: Normalize the file structure by eliminating internal repeating groups. Start with the outermost group, and create a new component file for each internal repeating group you have removed. The key for the new file is made up by concatenating the key of the old file and the key of the repeating group.

When this process is complete, we have an equivalent file structure made up of Normalized Private Component Files, shown below:

NPCF1	=	*Equivalent to PCF1*
	=	{Consignment-No. + Authorization-Code}
NPCF2	=	*Equivalent to PCF2*
	=	{Ship-Name + Consignment-No. + Agent-Name + Authorization-Code}
NPCF3	=	*Derived from PCF4*
	=	{Ship-Name}
NPCF4	=	*Derived from PCF4*
	=	{Ship-Name + Consignment-No. + Agent-Name + Consignment-Value}
NPCF5	=	*Derived from PCF4*
	=	{Ship-Name + Consignment-No. + Parcel-No. + Parcel-Value + Customs-Code}
NPCF6	=	*Equivalent to PCF5*
	=	{Ship-Name + Arrival-Date + Total-Value-on-Ship + Company-Name}
NPCF7	=	*Derived from PCF6*
	=	{Ship-Name + Arrival-Date + Total-Value-on-Ship + Company-Name} *DUPLICATE*
NPCF8	=	*Derived from PCF6*
	=	{Ship-Name + Consignment-No. + Consignment-Value}
NPCF9	=	*Derived from PCF6*
	=	{Ship-Name + Consignment-No. + Parcel-No. + Parcel-Value + Customs-Code} *DUPLICATE*
NPCF10	=	*Equivalent to PCF7*
	=	{Consignment-No. + Agent-Name}
NPCF11	=	*Equivalent to PCF8*
	=	{Warehouse-Location + Allocation-Indicator}
NPCF12	=	*Derived from PCF9*
	=	{Consignment-No. + Number-of-Spaces-Allocated}
NPCF13	=	*Derived from PCF9*
	=	{Consignment-No. + Warehouse-Location}
NPCF14	=	*Derived from PCF9*
	=	{Consignment-No. + Warehouse-Location + Parcel-No.}

As we continue decomposing the file structure into more and more single-minded pieces, some of the results begin to look a bit strange. NPCF3, to take one example, is an ordered list of names of all the ships on which information is maintained anywhere in the file structure. Since we are deriving the new file structure from the pattern of use of the old one, the fact that NPCF3 has emerged is an indication that one or more of the accesses needed to know the names of all the ships.

NPCF13 and 14 contain keys and nothing else. That may seem odd, but it only means that those files must be in "perfect sort" in order to perform their allotted functions. The function of such files is to keep track of correlations. (NPCF13 correlates consignments to warehouse locations where those consignments are stored. NPCF14 contains that same correlation and, in addition, it correlates warehouse locations to parcels contained in those locations.)

19.2.4 Further Decomposition

Only one further decomposition step is required before we can begin to reassemble the file structure. Our aim in this step is to derive an equivalent set of component files in which each one has a single purpose for existing. In order to proceed, we need some special definitions:

The *object of a file* is whatever the file records information about. (NPCF7 records information about ships; its object is ship.)

An *attribute* is an item of information recorded about an object. (Arrival-Date is an attribute of a ship.)

A *key attribute* is the principal identifier of an object. (Ship-Name is the key attribute of ship.)

A *correlation* is an indication that one object is associated with one or more other objects. (NPCF4 records a correlation between two objects — ship and consignment.)

What I have referred to above as the "single-mindedness" of a file can be expressed more exactly using these definitions. I consider a file to be single-minded if it records the attributes of one and only one object, or if it records a single correlation. The purpose of this fifth step is to decompose each file that does not qualify as single-minded into two or more files that are:

Fifth Step: Separate attributes from correlations. Associate attributes with the key attribute for the object they describe. Isolate correlations in pairs.

Any component file with more than one key attribute is a candidate for decomposition in this fashion. NPCF9 is a good example. It contains two correlations (ships to consignments on that ship, and consignments to parcels that make up the consignment). It also contains two non-key attributes (Parcel-Value and Customs-Code). When step 5 is applied to this file, we derive two pure *correlative files:*

NPCF5′   =   {Ship-Name + Consignment-No.}

NPCF5″   =   {Consignment-No. + Parcel-No.}

and one pure *attribute file:*

NPCF5‴   =   {Parcel-No. + Parcel-Value + Customs-Code}

We would have ended up with more than one attribute file if NPCF5 had contained non-key attributes of more than one object. I have tried to make the relationship between attributes and objects evident by the names given to data elements in this example. In a real-world situation, however, it is your application knowledge that helps you make the association. There is nothing about the construction of NPCF5 to tell you that Customs-Code is the code for a parcel. It is perfectly believable that it might apply to a whole consignment, or even to a whole ship. Any of those three possibilities could be correct, but only one is. In order to complete this step, you must determine the significance of each recorded data item and associate it with its proper object.

At the completion of step 5, the now fully decomposed file structure is the following set of Decomposed, Normalized, Private, Component Files:

DNPCF1   =   *Equivalent to NPCF1*
         =   {Consignment-No. + Authorization-Code}

DNPCF2   =   *Derived from NPCF2*
         =   {Ship-Name + Consignment-No.}

DNPCF3   =   *Derived from NPCF2*
         =   {Consignment-No. + Agent-Name + Authorization-Code}

DNPCF4   =   *Equivalent to NPCF3*
         =   {Ship-Name}

DNPCF5   =   *Derived from NPCF4*
         =   {Ship-Name + Consignment-No.} *DUPLICATE*

DNPCF6   =   *Derived from NPCF4*
         =   {Consignment-No. + Agent-Name + Consignment-Value}

DNPCF7   =   *Derived from NPCF5*
         =   {Ship-Name + Consignment-No.} *DUPLICATE*

DNPCF8   =   *Derived from NPCF5*
         =   {Consignment-No. + Parcel-No.}

DNPCF9   =   *Derived from NPCF5*
         =   {Parcel-No. + Parcel-Value + Customs-Code}

DNPCF10 =   *Equivalent to NPCF6*
         =   {Ship-Name + Arrival-Date +
             Total-Value-on-Ship + Company-Name}

DNPCF11 =   *Derived from NPCF8*
        =   {Ship-Name + Consignment-No.} *DUPLICATE*

DNPCF12 =   *Derived from NPCF8*
        =   {Consignment-No. + Consignment-Value}

DNPCF13 =   *Equivalent to NPCF10*
        =   {Consignment-No. + Agent-Name}

DNPCF14 =   *Equivalent to NPCF11*
        =   {Warehouse-Location + Allocation-Indicator}

DNPCF15 =   *Equivalent to NPCF12*
        =   {Consignment-No. + Number-of-Spaces-Allocated}

DNPCF16 =   *Equivalent to NPCF13*
        =   {Consignment-No. + Warehouse-Location}

DNPCF17 =   *Derived from NPCF14*
        =   {Consignment-No. + Warehouse-Location} *DUPLICATE*

DNPCF18 =   *Derived from NPCF14*
        =   {Warehouse-Location + Parcel-No.}

19.2.5 Combination of Component Files

Now at last we can begin to put the file structure back together again:

Sixth Step: Combine attributes of the same object into common component files (CCF).

Application of this sixth step results in the following equivalent file structure:

CCF1   =   *Composed of DNPCF1, 3, 6, 12, 13, and 15*
        =   {Consignment-No. + Authorization-Code + Agent-Name +
            Consignment-Value + Number-of-Spaces-Allocated}

CCF2   =   *Equivalent to DNPCF2*
        =   {Ship-Name + Consignment-No.}

CCF3   =   *Composed of DNPCF4 and 10*
        =   {Ship-Name + Arrival-Date +
            Total-Value-on-Ship + Company-Name}

CCF4   =   *Equivalent to DNPCF8*
        =   {Consignment-No. + Parcel-No.}

CCF5   =   *Equivalent to DNPCF9*
        =   {Parcel-No. + Parcel-Value + Customs-Code}

CCF6   =   *Equivalent to DNPCF14*
        =   {Warehouse-Location + Allocation-Indicator}

CCF7   =   *Equivalent to DNPCF16*
        =   {Consignment-No. + Warehouse-Location}

CCF8   =   *Equivalent to DNPCF18*
        =   {Warehouse-Location + Parcel-No.}

19.2.6 Common-Sense Simplification

Throughout the logical file derivation process, we have been replacing file complexities with processing complexity. The CCF equivalent file structure is free of most of the arbitrariness and redundancy that characterized the original file structure, but it requires more processing to make use of it. It may take dozens of accesses to the new file structure to accomplish what one access to the old file structure could do.

I make no claims for the efficiency of the evolving model. Its purpose is not to be efficient, only to be logical. Later, as we consider the performance of the new system file structure, we may want to insert selected physical traits to expedite file processing. But for now, our goal is to simplify the file structure as much as possible in order to come up with the most logical expression of it.

What other complexities could be removed by sacrificing file processing time? A likely candidate is any redundant sum such as the Number-of-Spaces-Allocated. Instead of storing that in CCF1, we could calculate it each time we needed it by looking through CCF7; all we have to do is count the entries there for the Consignment-No. in question. Similarly, we may be able to calculate Consignment-Value and Total-Value-on-Ship by summing the individual Parcel-Values, which are maintained in CCF5.

Seventh Step: Remove any stored data element that can be derived by reference to other stored data elements.

Each time this is done, there may be a corresponding simplification to the DFD. For example, Number-of-Spaces-Allocated no longer needs to be entered by Receiving. The simplification may occasion possibilities for repeated applications of any of the earlier steps.

Application of step 7 produces an equivalent file structure made up of Decomposed, Recomposed, Normalized, and Generally Logicalized Files (DRNGLF's):

DRNGLF1 = {<u>Consignment-No.</u> + Authorization-Code + Agent-Name}

DRNGLF2 = {<u>Ship-Name</u> + <u>Consignment-No.</u>}

DRNGLF3 = {<u>Ship-Name</u> + Arrival-Date + Company-Name}

DRNGLF4 = {<u>Consignment-No.</u> + <u>Parcel-No.</u>}

DRNGLF5 = {<u>Parcel-No.</u> + Parcel-Value + Customs-Code}

DRNGLF6 = {<u>Warehouse-Location</u> + Allocation-Indicator}

DRNGLF7 = {<u>Consignment-No.</u> + <u>Warehouse-Location</u>}

DRNGLF8 = {<u>Warehouse-Location</u> + <u>Parcel-No.</u>}

19.2.7 Packaging the File Structure

Look back over the set of files that now makes up our logical file structure. Notice that there are two very different kinds of files:

- *attribute files,* files that are accessed by a single key and contain all the attributes that pertain to that key

- *correlative files,* files that allow access among the files by correlating keys

DRNGLF1, DRNGLF3, DRNGLF5, and DRNGLF6 are clearly attribute files. They record all the information about consignments, ships, parcels, and warehouse spaces, respectively. The other files are correlative. They show us the correspondence between ships and consignments, for instance, or between consignments and locations in the warehouse.

You might observe that there is absolutely no redundancy over the set of attribute files. This always will be true if you have gone about the logical derivation properly. However, there is considerable redundancy in the correlative files. In fact, there is not one single data element in any correlative file that is not redundant! (If there were, it would be of no use in correlation.) The correlatives have information that is also in the attribute files. They also have internal redundancy. If there are 200 consignments in a given ship, for instance, DRNGLF6 will have to repeat the Ship-Name 200 times.

We would achieve Structured Analysis nirvana if we could somehow eliminate the correlative files; then we would have a totally non-redundant data set. It would be grouped nicely into four highly logical component files, each one accessible by a single key and detailing only those attributes that depend entirely on that key.

We cannot simply ignore the correlative files, though, because to do so would give us no way to link information together between attribute files. (As an example, we would have no way to complete logical access 6.) If we must implement the logical file structure with a set of simple files — sequential and/or direct access — then we will indeed require the correlative as well as the attribute files.

But the idea that our files must necessarily be simple sequential or direct is a physical consideration. If we are fortunate enough to have a data base processor at our disposal, we may be able to get *it* to implement the correlatives for us. The advantage of such an approach is that it shifts responsibility for control of the correlative redundancy onto the shoulders of the data base processor.

Since this is a very real possibility, we ought to provide for it by calling attention in our logical model to the difference between attributes and correlatives. We ought to recognize the very special nature of the correlative files: Each one is a linking mechanism, associating entries in one attribute file with entries in another. Then, if we do use a data base, the links will be implemented with a chaining feature. If we must use simple files, then there will be a correlative file for each link. In that case, use of the correlative file would be one physical implementation of the logical link.

Eighth and Final Step: Package the normalized file structure into a Data Structure Diagram, with one block for each attribute file and one logical pointer for each correlative.

The packaged equivalent file structure for our example is shown in Fig. 86, the completed logical file structure.

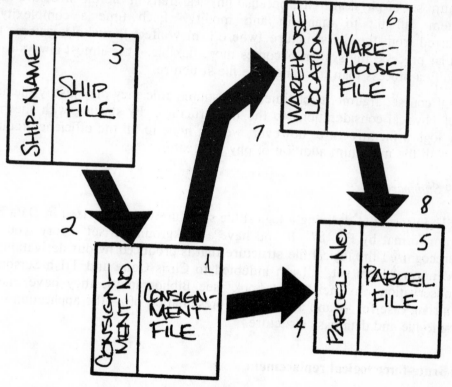

Consignment-File	=	*Was DRNGLF1*
	=	{Consignment-No. + Authorization-Code + Agent-Name}
Ship-File	=	*Was DRNGLF3*
	=	{Ship-Name + Arrival-Date}
Warehouse-File	=	*Was DRNGLF6*
	=	{Warehouse-Location + Allocation-Indicator}
Parcel-File	=	*Was DRNGFL5*
	=	{Parcel-No. + Declared-Value + Customs-Code}

Figure 86

19.2.8 What Have We Accomplished?

There may be any number of physical file structures that conform to a single logical equivalent. By identifying the most logical form of the file structure, we are sure to be starting from the best stepping-off place when we move into specification of the new system. Physical characteristics may be added later on, but none will be carried over unwittingly from the old system.

As the new system emerges, it may be useful to apply the logical derivation steps again. Our goal is for the new system file structure to have a minimum set of physical traits, because physical traits in the file structure make a system difficult to maintain and modify. Each time a complexity is transferred from the file structure (where it may affect the whole system) to a particular process, the system becomes more flexible. It is almost always easier to modify a process than to change the file structure.

Of course, arguments of time and resource efficiency may justify the addition of physical considerations to the file structure. By starting with the most purely logical file structure, however, we can hope to fill the efficiency requirement with the minimum addition of physical features.

19.2.9 Summary

The process of deriving a logical file structure is summarized in Data Flow Diagram format by Fig. 87. If you have a background in set theory, you may have recognized the logical file structure that is produced by our derivation as a "third normal form set." I am indebted to Chris Gane and Trish Sarson for this observation. In their latest book (see Bibliography), they have made a very strong case for — and an eloquent presentation of — the application of set theory to file and data base structure.

19.3 Brute-force logical replacement

Some of the physical characteristics of the current model can be eliminated by the mechanical process of expanding your DFD, and others can be eliminated by the cookbook procedure of logical file structure derivation. There are remaining physical traits, however, that defy treatment by either of these techniques. To cope with such problems, I have nothing more sophisticated to offer you than a brute-force approach. You have to pore over the details of your current model, looking for portions that are procedural, historical, political, or tool-related. Each time you detect such an offender, you have to replace it with its logical equivalent.

In order to detect physical aspects, ask yourself, Does this process *have* to work this way, or does it just *happen* to work this way? Does my model describe *what* the policy is, or *how* the policy is carried out?

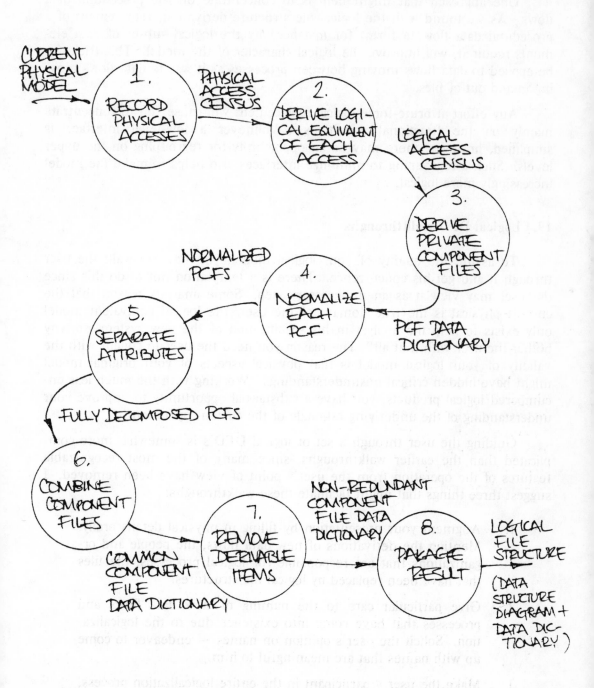

CURRENT PHYSICAL MODEL

1. RECORD PHYSICAL ACCESSES

PHYSICAL ACCESS CENSUS

2. DERIVE LOGICAL EQUIVALENT OF EACH ACCESS

LOGICAL ACCESS CENSUS

3. DERIVE PRIVATE COMPONENT FILES

PCF DATA DICTIONARY

4. NORMALIZE EACH PCF

NORMALIZED PCFS

5. SEPARATE ATTRIBUTES

FULLY DECOMPOSED PCFS

6. COMBINE COMPONENT FILES

COMMON COMPONENT FILE DATA DICTIONARY

7. REMOVE DERIVABLE ITEMS

NON-REDUNDANT COMPONENT FILE DATA DICTIONARY

8. PACKAGE RESULT

LOGICAL FILE STRUCTURE (DATA STRUCTURE DIAGRAM + DATA DICTIONARY)

Figure 87

One approach that might help is to concentrate on the procedural data flows. As we found with the logical file structure derivation, replacement of a procedural data flow (a form, for instance) by the logical subset of data elements required, will improve the logical character of the model. This trick can be applied to data flows moving between processes as it was to data flows moving in and out of files.

Any effort at brute-force logical replacement will cause you to concentrate mainly on the functional primitives. Whenever a low-level interface is simplified, however, there may be an opportunity for regrouping on the upper levels. Such a regrouping to minimize interfaces also helps to make the model increasingly more logical.

19.4 Logical DFD walkthroughs

To assure the quality of your logical model, you have to walk the user through it and get his concurrence. There is a temptation not to do this since the user may view it as an unnecessary step. Some analysts reason that the current physical is the true domain of the user; the logical equivalent model only exists to help the analyst in his specification of the new system, so why bother the user with it at all? The reason you need the user to concur with the validity of your logical model is that physical aspects of your original model might have hidden critical misunderstandings. Working with the much less encumbered logical products, you have a substantial opportunity to improve your understanding of the underlying rationale of the current operation.

Guiding the user through a set of logical DFD's is somewhat more complicated than the earlier walkthroughs, since many of the most recognizable features of the operation from the user's point of view have been removed. I suggest three things that might facilitate these walkthroughs:

1. Augment your presentation by filling in physical details orally. Mention the derivations of new data flows, the people and organizations that are responsible for logical processes, the files that have been replaced by logical file structures.

2. Give particular care to the naming of new data flows and processes that have come into existence due to the logicalization. Solicit the user's opinion on names — endeavor to come up with names that are meaningful to him.

3. Make the user a participant in the entire logicalization process, rather than doing it completely on your own and then showing him the result. By the time the walkthrough takes place, he should already be familiar with the product.

20 BUILDING A LOGICAL MODEL OF A FUTURE SYSTEM

The current-logical to new-logical transformation contains the true guts of the analysis phase. It is here that we build the logical model of the new system. Until now, all of our efforts have been dedicated to the study of current operations, *the way business is presently done.* Now we turn our minds to *the way business ought to be done,* i.e., the operation as it will be when our new system is completed.

For the first time in the analysis phase, we allow ourselves to consider the Charter for Change. This charter (documented as a product of the survey phase) is a statement of the rules that are to apply to the new system. Typically, they might call for automation of some set of procedures, or centralization, or decentralization, or putting some feature(s) on-line instead of batch, or adding a new line of business.

Our freight forwarding example might be made to serve as an illustration for this transformation, so we need a Charter for Change to apply to it. Here is one that will make for a sufficiently complicated test case:

> Due to the growth in volume of freight being forwarded and the attendant growth in the cost of related paperwork, the decision has been made to automate certain features of information processing in support of the operation. In particular, it is anticipated that the major files will be on-line, with a simple history file of paper documents to serve as a fallback medium and archive. This means that terminals currently used for the NCSS time-sharing system will now have access to our own system. The new system will do space allocation (as the old NCSS did), and support inquiry, removal, data entry of new arrivals, as well as generation of Customs Department and IRS reports.

As a work description, that is splendidly vague. But the Charter for Change must allow considerable leeway in the degree of automation and amount of system function, since the complete cost-benefit analysis is still to come. The eventual system which will result from our project might be a simple replacement of the current warehouse space allocation system with some rudimentary library-type function to support the additional file requirement. On the other hand, it might be something considerably more grandiose; it might involve tracking and timing of consignments through the warehouse, on-line management of the conveyor network, cost accounting on handling expenses, and more.

As a result of previous analysis phase efforts, we now have a completed logical model (Data Flow Diagram, Data Dictionary, Data Structure Diagram, and mini-specs) of the current system. When the current logical to new logical transformation is over, we shall have a similar logical model describing the new system. In each case, the term "system" refers to a connected set of procedures, some manual and some automated. Logical models do not distinguish between automated and manual procedures.

How to begin modeling the new system? Of course, you could simply take out a fresh piece of paper and start over. That would involve coming up with a new Context Diagram and then performing a top-down partitioning with Data Flow Diagrams, Data Dictionary, etc. There is, however, a slightly easier way to get started. It involves salvaging a portion of the current logical model, and using that as a base upon which to build.

20.1 The Domain of Change

If you have been suitably defensive in the original selection of a context for your study of the current environment, then it will include substantially more than the area likely to be affected by the change specified in your charter. The implication of that is that part of your new logical model will be exactly the same as the corresponding part of the old logical. If you can isolate this part and adapt it, you will have a useful headstart in constructing the new logical model.

With reference to your old logical DFD, the Domain of Change is defined as the union of all parts that *must be or are likely to be changed* by the new way of doing business. The Domain of Change is the part of the old logical model that will have to be thrown away (we have to rework that part). The "Domain of Non-Change" is the part that can be retained.

In order to establish the Domain of Change, go over the current logical DFD, bubble by bubble. Work at the bottom level. For each functional primitive, ask, Is anything about this process likely to be different in the new way of doing business? Might it be automated or partially automated or eliminated or done in a different order or different way? If so, it is part of the Domain of Change. If not, if the process is totally unaffected by even the most sweeping changes the charter authorizes you to consider, then the process lies outside the Domain of Change. Proceeding in this fashion, you can separate the current logical DFD into two domains: the affected and the unaffected.

Once you have identified the Domain of Change and marked it on your Data Flow Diagram, your result is in the form of Fig. 88. Do not be concerned if you find that it has not one but several Domains of Change, totally disconnected from each other. That is just a consequence of the way your Data Flow Diagram is laid out on the paper — it has no additional significance.

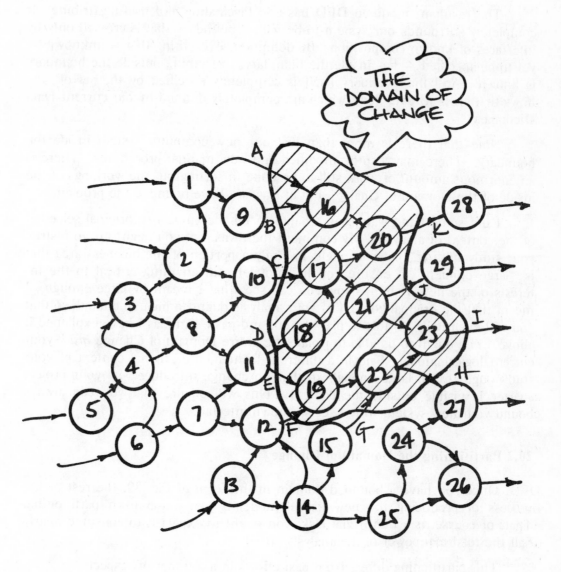

Figure 88

Now I propose that you go into cut-and-paste mode. (Analysts are super cutters and pasters.) Cut out the entire Domain of Change and lay it aside. Replace it with a single bubble labeled Domain of Change and connect the interfaces. That should give you a working document like the one shown in Fig. 89.

The resultant modified DFD has one fascinating and useful attribute: It completely surrounds our system-to-be with a boundary that is crossed only by interfaces of known composition. By definition, everything that is unknown — yet to be specified — lies inside the boundary; everything outside the boundary is known. And the boundary itself is completely specified by the set of data flows that cross it, data flows which are completely defined by our current Data Dictionary.

Note that there is more than just our new computer system inside the boundary. There may also be modified and new manual procedures. There is an enormous amount of work still to be done in sorting out the workings of the inside of the Domain of Change, but at least we have bounded the problem.

I did some hand-waving in Chapter 5 on the subject of original selection of the context of analysis. My comment then was that you ought not to restrict your study unduly, that you ought to study a portion of the business area that is "large enough" to ensure you had not left out anything critical to the interests of the new system. Now you can see what I meant by large enough. I meant that the original context of your analysis ought to include enough so that you will eventually be able to play the cut-and-paste game as I have explained it above. You ought to be able to isolate the entire Domain of Change *inside* your current logical model. Said in a different fashion, the original context of your study ought to be large enough so that no change introduced by your project charter is visible outside the context. This will ensure that you can always bound your new system with a set of known interfaces.

20.2 Partitioning the Domain of Change

Once you have a bounded domain in the form of Fig. 89, the rest of the process of producing the new logical model is just a top-down partitioning. That, of course, is a non-trivial task; you ought to expect it to consume nearly half the total manpower of the analysis phase.

This partitioning differs from past efforts in a number of respects:

- It is truly top-down. Since the top is known (the new Context Diagram is effectively the Domain of Change with its input and output data flows), there is no need to start in the middle as is sometimes required in studying the current physical environment.

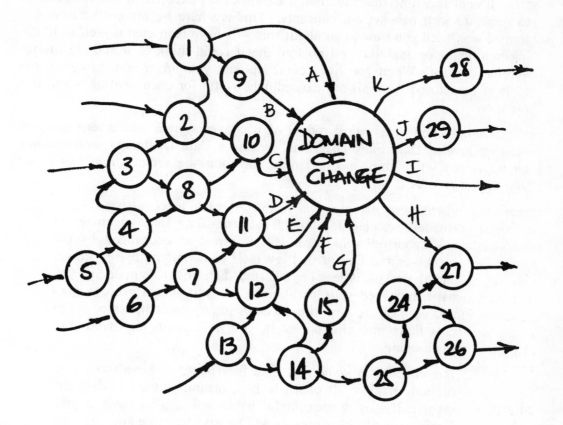

Figure 89

- The analyst is boss. Instead of being guided by the user's political and procedural view, the analyst partitions according to his own standard: minimization of interfaces.

- There is a new set of rules in force. I refer to the Charter for Change. It may call for added function, new processes and data flows, increased accountability, etc.

It is at this time that the analyst exercises his experience and imagination to come up with new system concepts. This is where he *invents* the new system. I won't tell you how to go about this — I have restricted myself to teaching new tools for analysis, and no tool that I could think of would aid the invention process. When you have come up with your invention, however, the tools of Structured Analysis are exceedingly useful for documenting it and trying it out.

Construction of a logical model of whatever system you invent involves drawing leveled Data Flow Diagrams, writing a Data Dictionary, and creating an integrated set of mini-specs. I offer the following set of guidelines to aid you in this process:

1. *Partition to minimize interfaces.* At each level, strive for the smallest possible amount of data transfer. Take advantage of any opportunity to show less information exchanged by the processes and files at that level. If you combine two processes and the result is less complicated than either component, you have improved your partitioning. If you split a process apart and each of the new subprocesses has a less complicated set of data flows into and out of it, you have again improved your partitioning.

2. *Pay attention to names.* A data flow ought to have some conceptual integrity: It ought to be a meaningful set of data elements. If it is, a meaningful name will suggest itself to you. Useful, functional processes will be easy to name and they will have concise, strong names. You should never need to use such wishy-washy words as process, handle, data, input, and output. Don't struggle to come up with a meaningful name for a process or data flow. If one doesn't suggest itself readily, that means your partitioning was not so hot. Go back and think it out again.

3. *Respect conceptual counting limits.* Small sets (seven or fewer elements) are much easier for the human mind to deal with than larger ones. Sometimes, there will be an overriding reason to portray more than seven processes on a given figure (if, for instance, there is a natural parallelism of data that cries out to be respected). Other times, however, the extent of partitioning at each level is largely left up to you. By all means, restrict yourself to seven or fewer processes per figure.

4. *Be complete.* Make sure that each and every data flow is specified. The one you forget to name is invariably the one that defies naming. Make sure that all figures balance with their parents.

5. *Respect the data conservation rule.* Any data element that flows out of a process must flow into it in some form (unless it is absolutely fixed). Conservation errors mean that some interface has been left out. When you add in the required input, you may find that the partitioning needs to be thought out again.

6. *Use the old logical model to guide you.* Be suspicious, however. Concepts that were fine in the old environment may not be acceptable under the new set of rules.

7. *Use the Data Dictionary to keep track.* Each time you create a new data flow, be sure to define it. If you can't define it, it may contain a multitude of sins — enough to invalidate your partitioning.

8. Most important of all, *be prepared to start over.* And over and over and over. Be iterative. Use a lot of paper. Keep trying out new partitioning ideas in the hopes of coming up with a version that is even less complicated, and more readable at every level, than your current best.

20.3 Testing the new logical specification

Your logical description is complete when you have a coherent balanced set of DFD's with all supporting documentation. In order to verify its meaningfulness, you must now show it to the user and help him to understand its significance. It is important that you both think of the new logical descriptions as a true model of the new environment. Just as an architect walks his client through a physical model of his home-to-be, you have to guide the user through his simulated new environment. Take the roof off for him and let him look around. Encourage him to imagine himself and his workers pushing data through the new system. Make up sample inputs and outputs for him from the Data Dictionary definitions.

Urge the user to help you "debug" the model while it is still on paper. You may observe that you have a certain mental block against finding flaws in the model since you have so much work invested in it. The user has no such problem. His attitude at this time is the perfect one for debugging. Each time he can find something wrong, he is exhilarated. *And* he has managed to save himself some grief downstream. Don't be exasperated by his obvious relish when he does find a flaw. Fix up the model. Be willing, too, to tailor it to his particular wishes, if possible. For every ten arbitrary changes he imposes on you, he will save your life with one important fix. This is not the time to ask the user to be lenient. On the contrary, he ought to be allowed some whims while the system is still made only of paper. Now is the time to co-opt him, to make the model so acceptable to him that it becomes his own. Your goal is to make him feel that, by imposing his modifications on the model, he is establishing the eventual shape of the system. This will give him a sizable measure of responsibility for the system when it is delivered, and will make him feel every deficiency is at least partly his fault. That frame of mind makes him doubly helpful during analysis and more than normally docile at acceptance time.

At this stage, the concurrence that you need to solicit from your user consists of statements such as: Yes, I could do business the way this model portrays. No, I can't think of any better way to do business, given the constraints imposed upon us.

21 PHYSICAL MODELS

The last transformation calls for incorporating a minimum of physical information, principally the man-machine boundary. Since the logical model does not distinguish between automated and manual procedure, it lacks an important quality that any Target Document requires: determination of the scope of the subsequent development effort. Only when you have established which of the primitives of your model will be performed by a machine and which by humans will you have completed the specification of your project targets.

The scope of automation is determined by a cost-benefit study. You cannot state definitively how large a development effort is justified until that study is complete. So the cost-benefit study is a key part of the new-logical to new-physical transformation.

The transformation proceeds in the following manner:

1.   Prepare one selection option by setting out a tentative man-machine boundary.

2.   Add any implementation-dependent features.

3.   Repeat steps one and two until enough options have been prepared.

4.   Select the best option.

21.1 Establishing options

Please look at the pictures in Figs. 90, 91, and 92. (I promise to leave out 3,000 words.) Let Fig. 90 represent your final logical Data Flow Diagram, portrayed at the bottom level. Then Figs. 91 and 92 might represent two options developed as part of the logical to physical transformation. They differ only in degree of automation. Fig. 91 might be the "big machine solution." It calls for practically everything to be done inside the machine; almost none of the logical procedure remains manual. It can be expected to be more expensive than the second option, shown in Fig. 92, but it will have some added benefit. Fig. 92 might be the "minicomputer solution." It would leave substantially more of the work in the hands of humans, but it would also cost less to install.

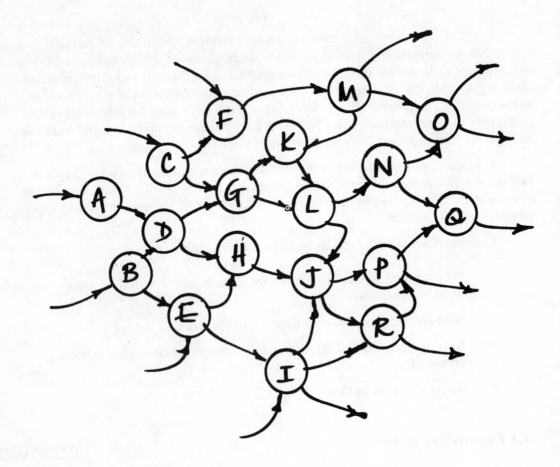

Figure 90

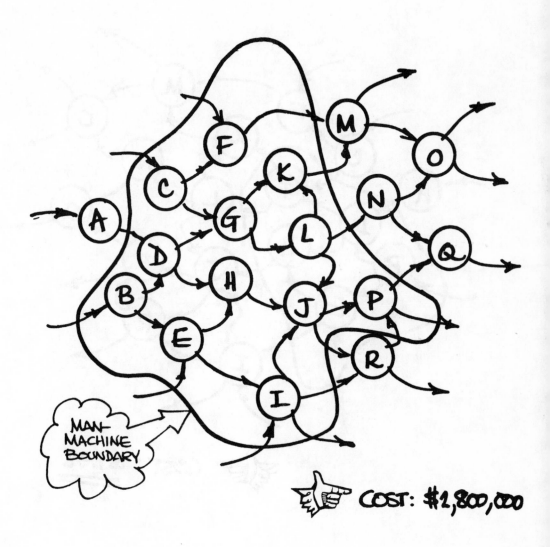

COST: $2,800,000

Figure 91

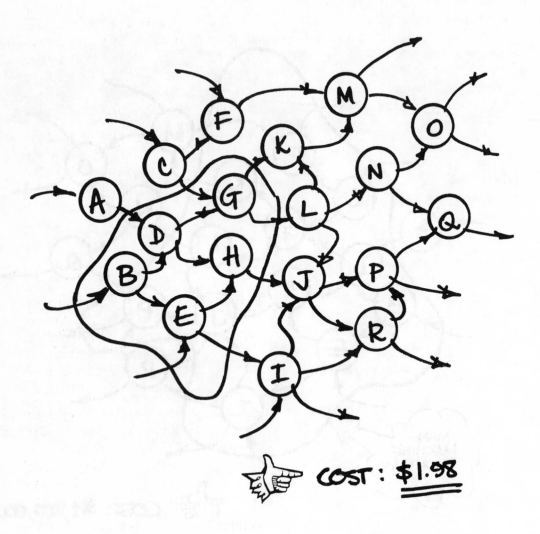

COST : $1.98

Figure 92

Your goal is to come up with options, all of which should be acceptable. Of course, only one will be selected, but none of them can be a priori unacceptable — if it were, it would not be a true option.

I promised not to get into the subject of politics, but this is a political situation that cannot be ignored. (In other words, I am breaking my promise.) The analyst can create an untenable situation for himself by selecting a single option and then presenting it as the only possibility. This means that, forever after, the system will be thought of as *his system*, the analyst's folly. Since management and user staff played no part in deciding how big a system to build, if it is either too big or too small, the analyst is clearly at fault.

This situation can be defused by creating a meaningful menu of viable alternatives, and then allowing management and the user to participate in the selection. Their part in the selection process puts the shoe on the other foot. Now it is *their system*. If it is too small to be useful or too large to be installed, that is *their failure* as well as the analyst's. This feeling of shared responsibility is essential to successful completion of the project.

21.2 Adding configuration-dependent features

It is highly unlikely that the only difference between a minicomputer solution and a maxicomputer solution would be which processes each solution automated. In general, the more expensive solution would do considerably more; it might implement functions that the smaller solution did not accomplish in any way. I refer to such functions as configuration-dependent features. Each of your options might have one or more configuration-dependent features.

When you add new features, you are back in inventing mode, doing again the work of logical modeling that was part of the last transformation. In fact, you might think of these two transformations together, since you must do a little of one and then a little of the other. The final result is not a single model but several models, one per option. Each model is highly logical: The only physical information in it is its man-machine boundary.

When the configuration-dependent features are added, you end up with a set of annotated DFD's like the one in Fig. 93. The annotations consist of incorporated configuration-dependent processes (shown here as solid bubbles) with their required data flows.

21.3 Selecting an option

This is where you perform your cost-benefit study on each of the options. As I said earlier, I will not enter into the "hows" of cost-benefit analysis since Structured Analysis does not offer additional insight to this process. Cost-benefit analysis proceeds in the structured environment pretty much as it did in the more classical environment.

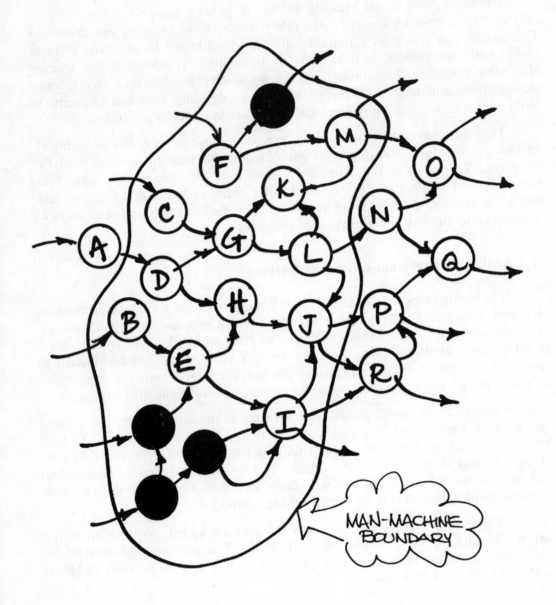

Figure 93

Regardless of how you go about cost-benefit analysis, it is important not to do it before you have a workable set of models. Management would like to see the cost-benefit study completed during the survey phase. (Why spend all that money on the analysis if the cost-benefit doesn't warrant it?) But the sober fact is that you cannot analyze the trade-offs until you have something to analyze. The idea of performing an early cost-benefit analysis is largely a fiction.

When each of the options has been quantified in terms of all its varied costs and the attendant benefits, selection can take place. The same comment about the dangers of doing the cost-benefit study too early applies to the dangers of selecting the scope of automation before this point in the life cycle. I know, however, that this happens all the time. Machines are purchased and key decisions are made before any analysis has begun. If you are the analyst on a project that works like this, where the boss runs out on Day One and buys a brand-new sparkling DataSchwartz 2938 computer with the DS2 Multivariant Deeply Differential Operating System — well, you have my sympathy. Relax and enjoy yourself. You will not be able to conduct your project exactly as I have advised, and it may not work out too well in the end. But maybe you will want to write a book on analysis one day, and then the experience will be invaluable.

22 PACKAGING THE STRUCTURED SPECIFICATION

All the conceptual thinking of the analysis phase is now complete. The only remaining task is to collect the products of analysis and organize them into the finished Structured Specification. This collecting and organizing process is called "packaging."

The packaging effort may involve any of the following:

- redrafting to call attention to key interfaces

- preparing a Structured Specification guide to help readers

- preparing supplementary material to augment the Structured Specification

- filling in details that had been deferred until now

Since it is the most substantive, I begin with the last of these.

22.1 Filling in deferred details

Throughout the earlier part of analysis, I urged you to defer specifying certain details about the new system. For the most part they were physical considerations such as formats, layouts, and internal system structure, items that should be considered much later in the life cycle. Others of the deferred items, however, ought to be picked up now and included in the Structured Specification.

The techniques involved in dealing with these last minute details are not new to you. I suggest you use the tools of Structured Analysis to include them in the specification. To guide you in this process, the following is a list of details that were left until now to be specified:

1. *Error messages.* Existence of the error paths was established earlier. Now you must go back and decide the exact text of each message.

2. *Startup and closedown.* The model is concerned principally with the steady state. The Structured Specification, on the other hand, must show how the steady state is attained and how it

can be terminated. This can be done with a marked up set of Data Flow Diagrams and additional Data Dictionary entries.

3.   *User-sensitive control information.* Your Data Flow Diagrams, quite properly, do not contain any control paths or flow of controlling information. While you can construct a meaningful model without showing any control information, sometimes the user's concern can be set to rest by adding a small number of controls on certain key flows. I suggest that you do this now; do it in the form of annotations to your Data Flow Diagrams — dotted lines, for instance, with notes to indicate the meanings of these added constraints.

4.   *User-sensitive formats.* Sometimes the format of a given report is so important to the user that he will not let you terminate the analysis phase until the format is completely specified. This may happen with an output form that is subsequently used as input to another system, or one whose format is determined by law. In such cases, users are justified in insisting that the actual format be included in the Structured Specification.

5.   *Conversion.* I have not mentioned this important topic until now. In the simplest cases, the analyst can leave conversion worries until the packaging stage. But there are important exceptions. There are projects that require construction of an entire conversion system to serve (on a one-time basis) the conversion process. Clearly that kind of packaging requirement ought to be dealt with much earlier in analysis — in fact, the models constructed during the analysis phase ought to include a model of the conversion system.

6.   *Performance.* While the system's expected performance has molded thinking throughout the analysis phase, no performance information has yet been incorporated into the Structured Specification. This should be done now. A statement about performance targets should be drafted and added to the Structured Specification. If possible, it should take advantage of the PERT-like nature of the DFD's to portray critical path timing restrictions directly on the DFD network.

There may be other details that should be added at this time, information about true project targets. But it is worth being rather defensive about this in order to avoid burdening the specification with decisions that ought to be left to the discretion of the designers. Some analysts are eager to design all the input and output formats and impose those formats as requirements on the implementation team. This constitutes overspecification in my book. The user may be utterly indifferent to the format of a particular report. Once he has verified that its content (documented by the Data Dictionary) is acceptable, he could care less where the various items end up on a page. But the designer may care desperately. There may be an order of magnitude difference in the implementation effort for two different report formats, both containing the same information. Since only the implementors can judge which is preferable, we ought to leave the choice to them.

22.2 Presenting key interfaces

As part of the packaging task, I suggest that you now regroup and redraft your Data Flow Diagrams in such a way that all the key interfaces are expressed at the very top level. In particular, the man-machine interface ought to appear there; this implies that the automated portion of your system should be represented as a single bubble on the top-level figure.

Fig. 94 is an example of the kind of packaged result I advocate. It shows the top-level Data Flow Diagram as it might look for a system to be installed in the freight forwarding operation we looked at earlier. The man-machine interface is the set of data flows that move into or out of Bubble 5. Bubble 5 together with its lower-level figures (all the way to the bottom) plus its associated Data Dictionary entries and mini-specifications constitute the specification of the automated system being implemented. The rest of the top-level processes together with their lower levels and supporting documentation complete the Structured Specification. (The Structured Specification consists of a description of the *whole* system, not just the automated portion of it.)

By forcing the man-machine interface to the top, readability of the Structured Specification is enhanced. This approach also affords you the possibility of declaring key political interfaces that are relevant to the system. In Fig. 94, I have gone back to the organizational groupings (Front Office, Receiving, Warehouse, and so forth) that were used in the earlier physical DFD's. All of this helps to make the top level of the Structured Specification a usable picture of the organization as it will appear when the new system is installed.

In the case of a distributed system, you might choose to show the machine-machine interfaces at the top level as well. Fig. 95 depicts the top level of a banking application in which a mini-machine in the branch and a larger machine in the back office combine to do the complete automated function. The man-machine interface in this example is the set of data flows communicating between the automated portion of the system (Bubbles 1 and 2) and the manual portion of the system (all the others). The machine-machine interface consists of the data flows passing between Bubbles 1 and 2.

22.3 A guide to the Structured Specification

Since the Structured Specification is something of a newcomer to most organizations, you should provide a guide containing introductory and tutorial material to explain the conventions that govern the use of the Structured

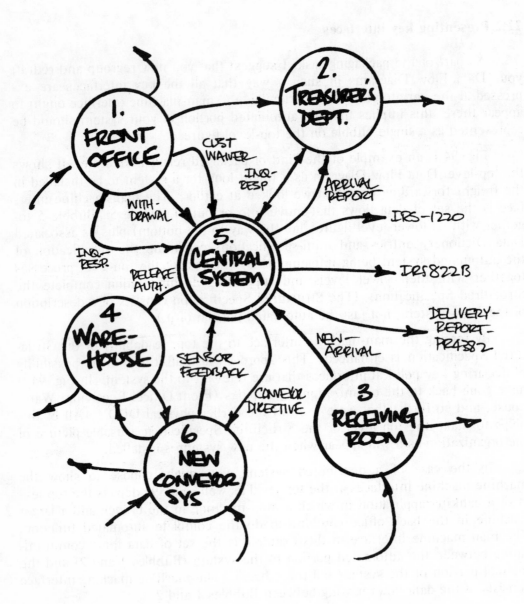

Figure 94

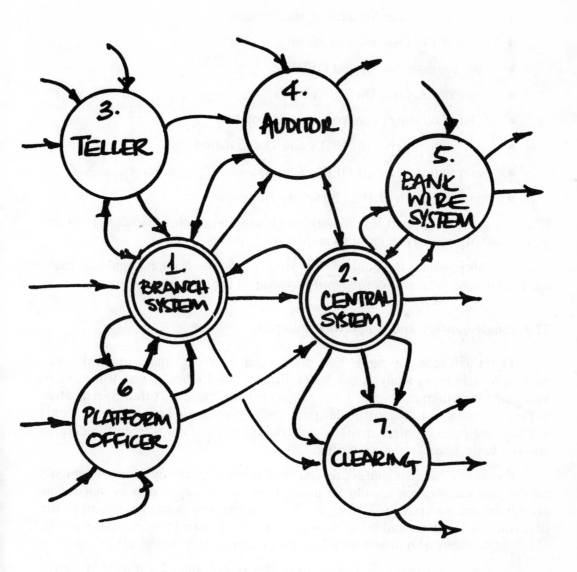

Figure 95

Specification. Whatever form this guide takes, it should be substantially the same from one specification to another — once you have written it for the first, the same document should suffice for those that follow.

The guide to the Structured Specification should cover all of the following:

- elements of the Structured Specification
- role of the Data Flow Diagram
- conventions for leveled DFD's
- role of the Data Dictionary
- Data Dictionary conventions and notation
- correlation between DFD's and Data Dictionary
- correlation between DFD's and mini-specification set
- interpretation of Data Structure Diagrams

It makes sense to support this material with some examples, particularly examples of Data Dictionary definitions and their meanings.

If at all possible, the guide to the Structured Specification should be packaged separately, with a reference to it included in the specification proper.

22.4 Supplementary and supporting material

There will probably never be a project that is so "normal" that any cookbook approach to its analysis can be sufficient. Undoubtedly you will have to augment the minimum recipe I have provided for the Structured Specification in order to tailor it to the particular needs of any project. These enhancements will take the form of supplementary material included along with and integrated into the basic Structured Specification.

Further, I have to confess to you that in spite of all my pronouncements on the way specification should be done, most of the organizations that follow my advice deviate from it somewhat. They invent new notations to add to the fundamental DFD set and adopt new conventions to suit their particular needs. (I have one client who insists on filling his diagrams with triangles!)

To the extent that you deviate from the specification formula that I have espoused, please keep this idea in mind: *Beware of redundancy and overspecification — these are the analyst's enemy.* It might be convenient to have a little hierarchy listing that shows what all the processes and subprocesses are. But that listing is redundant; a cross-reference table that correlates data elements to processes or to superordinate data flows is also redundant. When you add such a nifty feature to your specification, you may be creating an additional maintenance problem. Whenever you add something to the specification which will intrude upon the degree of freedom that ought to be left to the designer, you risk increasing the cost of implementation. The more freedom you can allow the designers to maneuver within your real requirements, the more they can take advantage of opportunities to control costs and to maximize the maintainability of the system.

At packaging time, you may want to introduce some examples to help your readers understand the new environment. Make sure that these examples are *not* an integral part of the Structured Specification, and that everyone understands they will not be kept up-to-date.

PART 6

STRUCTURED ANALYSIS FOR
THE LATER PROJECT PHASES

23 LOOKING AHEAD TO THE LATER PROJECT PHASES

The analyst's function does not end with completion of the analysis phase. He fills a continuing need during the rest of the project, participating in the design and implementation, acceptance testing, parallel testing, conversion, and the early operational period. Because of his broad knowledge of the environment and the requirement, there is usually great demand for the analyst to take on new responsibility as the project progresses. (There may be great demand for the analyst to take on new responsibilty in some other company, but I won't go into that.) He may be called upon to become the Project Manager, for instance, or the Chief Programmer. But the role of the analyst proper also carries on into the succeeding phases of the project. That continuing analysis function and the ongoing application of Structured Analysis tools and principles are the subject of the remaining chapters.

23.1 Analyst roles during design and implementation

At the conclusion of the analysis phase, project targets should have been selected and documented. But it is unreasonable to expect that those targets will remain absolutely constant during the rest of the project. Particularly if the development period is a long one, you should expect the targets to change rather substantially due to the following:

- corrections imposed by the user as he comes to understand better the significance of the system-to-be

- changes imposed by the development team to reduce the scope of effort

- changes imposed by management (modification of the project charter)

- new requirements imposed by the user to reflect the continuing evolution of his operation.

The most important of the analyst's ongoing responsibilities is, therefore, the careful tracking of the moving project target. He has to maintain an open dialogue with the user staff and make sure that each and every modification of the target causes a corresponding update to the Target Document.

Simultaneously with his monitoring of the target, the analyst must monitor and guide the development project that is trying to hit the target. He has to provide feedback to the designers when their tentative efforts fail to meet the requirement. He must be able to judge the applicability of the design and its quality. He cannot approach this role solely as a critic; he must be prepared to take an active part in the design to keep it on target.

Because it is the analyst's task to coordinate the moving target and the development effort that is proceeding toward that target, he may become the person to establish and enforce a development methodology. He may also be called upon to provide key estimates for the development effort and checkpoints to measure progress of that effort against the plan. In fact, any of the parameters that could affect project performance may become the de facto responsibility of the analyst.

Finally, the analyst is the natural candidate to take charge of the acceptance testing process. Development and administration of the tests are his particular responsibility, especially since these require a renewed and vigorous interface with the user.

In summary, the role of the analyst during the post-analysis phases is likely to include all of the following tasks:

- ongoing specification and respecification
- active participation in the design process
- selection of tools and methods for development
- project planning, estimating, and control
- acceptance testing

We will look at each of these areas, but first a comment on a particular problem that falls to analysts who have used Structured Analysis techniques.

23.2 Bridging the gap from analysis to design

Because Structured Analysis is a new approach to the specification process, a question naturally arises about how the products of this modified project phase are used to provide a smooth transition to design. The answer to that question has to do with the way the design phase is administered and, in particular, with the set of techniques and strategies that has come to be known as Structured Design. Briefly, there is a natural relationship between the Structured Specification and the early activities of Structured Design. In fact, the concept of the Structured Specification was expressly tailored to provide the cleanest possible interface to the design process.

One of the strategies for starting off design is called *Transform Analysis*. Transform Analysis is a cookbook strategy for deriving a design (usually portrayed by a Structure Chart) directly from the Data Flow Diagram. The result is not a finished design — it requires substantial reworking before the designers will find it acceptable — but it does provide a good beginning to the design process. The rest of that process is primarily concerned with evaluation and refinement of the initial design.

Since the analyst must participate in design, in either an active or a passive manner, he ought to have a good understanding of the techniques of Structured Design. With this goal in mind, I have compiled a short description of these techniques and their underlying philosophies and presented it in Chapter 25. The scope of that description is necessarily limited, but references are provided for each of the major topics to assist you in pursuing them further.

23.3 User roles during the later phases

The user, too, has a continuing role in the post-analysis phases of the project. He has to keep the project team posted on changes in the business environment. He must take an active part in acceptance testing. A particularly active and concerned user will also keep an eye on the detailed work of design and implementation to ensure that his requirements have been clearly understood.

Close participation of the user during development can be an enormous boon to a project. In spite of our best efforts, no specification can ever be 100 percent complete, nor can it be completely accurate. Constant interaction with the user is therefore necessary to complement the written statement of requirement. But users are often reluctant to be involved so deeply in the technical development because of the foreignness of the work (and, perhaps, of some of the people doing the work). It is up to the analyst to create a workable interchange of information between the user and the development team. He has to look upon the user as a valuable expert in subjects essential to the project, but an expert who happens not to speak the language. The analyst must serve as his translator in order for the user to be able to help. With the aid of the analyst, the user can work as an integrated member of the project team. Without the analyst's translation service, communication breaks down and the user participates less and less.

So the analyst continues to participate in the development phases, serving as specifier, design consultant, acceptance test administrator, project planner, controller, and translator. With this much introduction, let's go back and look in detail at each of the analyst's ongoing responsibilities.

24 MAINTAINING THE STRUCTURED SPECIFICATION

System analysis continues beyond the end of the analysis phase. The major analytic activities — user dialogue, modeling, documentation of development targets — can be expected to last the entire life of the new system. Those activities, when they were performed during the analysis phase, contributed to generation of the Structured Specification. When they are performed during the later project phases, they cause the Structured Specification to be updated.

The concept of the Structured Specification was derived with the goal of maintainability in mind. Classical Target Documents were impossible to update as the requirements changed; the Structured Specification is a new kind of Target Document intended to remedy that, intended to support tracking of moving project targets. In this chapter we shall examine the techniques and procedures for keeping the Structured Specification up-to-date.

24.1 Goals for specification maintenance

What goals shall we adopt for our specification maintenance procedures? Of course we would like to minimize the cost of processing each proposed change, from beginning to end. The total cost allocated to a given change has to include all of the following:

1. *Formulation of the change request:*[1] Prepare a concise statement of the change being considered.

2. *Evaluation of impact:* Determine all the attendant costs and benefits.

3. *Decision:* Accept, defer, or deny the change request.

4. *Incorporation of the change:* Update the Structured Specification to reflect the modified requirement.

5. *Implementation of the change:* Redirect the development effort to take the modified requirement into account.

---

[1]The term "change request," as I have used it here, refers to a formal proposal to alter the Structured Specification and redirect development toward a modified target.

Normally, more changes are proposed than are actually implemented. This implies that the process of formulating new change requests and evaluating them has to be particularly efficient. In fact, our specification maintenance procedure ought to concentrate on how the changes are formulated, since that has a strong effect on each of the subsequent steps.

Formulation of change requests is very like the analysis phase modeling activity. The major difference is that during the analysis phase the entire system was being modeled, and now it is only an increment to the system that must be modeled. The requirements that applied to our earlier modeling still apply: The results of the process have to be graphic, partitioned, iterative, rigorous, and non-redundant.

Iteration is perhaps the most important quality in maintenance mode. With the development project already under way, the expected impact of each proposed change is magnified. Not only must we consider the cost of adding the change itself, but also the disruptive effect that adding it may have on the rest of the development effort. There are more people involved now in evaluation of the model. The importance of a short author-reader cycle increases with each new person who has to review the work. So, however we elect to formulate the change request, we must be able to move the request quickly back and forth over the desks of all the interested parties until it has been put into a form that is acceptable to all.

The incremental model must fit readily into the context of the current Structured Specification. This insures that the cost of updating that document, if and when the change is accepted, will be minimal. More important than that, it helps reduce the cost of evaluating the change. Since the design derives its shape from the shape of the model contained in the Structured Specification, the more easily each change can be correlated to the system model, the easier it will be to determine which modules and programs are likely to be affected.

Finally, it is desirable for the skills and methods required of the analyst during the specification maintenance period to be the same as, or similar to, the skills and methods he has used earlier.

In summary, goals for the specification maintenance procedures are the following:

- Each change request should be formulated as an incremental model.

- The change request must have a well-specified interface to the overall system model contained in the Structured Specification.

- The change request ought to have all the qualities that we expect of the Structured Specification itself; in particular, it should be iterative.

- The change request should be easily related to the evolving design.

- The analysis skills and technologies of the maintenance period ought to be no different from those already mastered during the analysis phase.

24.2 The concept of the specification increment

You can probably guess what's coming next. I advocate that the familiar tools of Structured Analysis be used during specification maintenance pretty much the way they were used during the analysis phase. That means you conduct your dialogue with the user (at this stage, a dialogue concerning change requests) as you conducted it from the beginning of the project: After each interview you construct a model (Data Flow Diagrams, Data Dictionary, and so forth) of what the user has described to you. You show that to the user and encourage him to correct and refine it through a series of iterations until it is right.

The model you build to describe a given change request is what I referred to earlier as an "incremental model." It is so named because it describes an increment to the system or, more precisely, an increment to the system model contained in the current version of the Structured Specification. If the change request is accepted, that model will be revised. The revised system model will consist of the old model plus the increment, which may possibly replace or eliminate some existing portion of the old model. I am suggesting that this incremental model ought to be your formal statement of the requested change.

The change is packaged together with some indication of how it fits into the existing system model. It is common to add certain overhead items to this package:

- change request number
- requestor's name
- short narrative description of the change
- reason for the change

The final result (incremental model plus its relationship to the Structured Specification model plus overhead) is called the *Specification Increment Document* or SID. There is one SID for each requested change.

As an example, suppose you are midway into the design phase of a secure system to support armed forces logistics (whatever those are). You might encounter a change request such as this one:

==================================

CHANGE REQUEST FORM

==================================

CHANGE REQUEST NO.: 12
REQUESTED BY: Col. Cathcart

ABSTRACT: Add a feature to report each night on
             all denied read requests.

RATIONALE: Since any file security system can be
            penetrated if given sufficient opportunity,
            it is important to have some indication of
            each frustrated attempt.

PROCESSES AFFECTED: 1.4.2
DATA FLOWS AFFECTED: Security-Violations-Report
FILES AFFECTED: Anomaly-File
APPROVED: Gen. Dreedle

==================================

The complete SID would be made up of this Change Request Form, a revised Data Flow Diagram for Diagram 1.4.2, and incremental Data Dictionary
(both shown in Fig. 96), plus revised Structured English descriptions of the
modified processes. Those people called upon to evaluate the change would be
able to integrate the SID temporarily into their copies of the Structured
Specification by replacing the affected portions by components of the SID.

Using this approach, each change request would be considered individual-
ly, as though it were the only change being considered. Sometimes changes
must be considered in groups. When this is the case, it is best to make a combined SID or at least to emphasize the linked nature of the several separately
submitted SID's.

In general, the composition of an SID is

1.   a Change Request Form of some sort to present overview and
      control information and to document the processes, data flows,
      and files affected by the proposed change

2.   annotated Data Flow Diagrams to replace each affected Data
      Flow Diagram in the Structured Specification

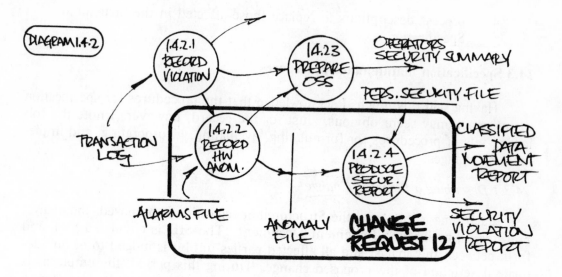

Incremental Data Dictionary

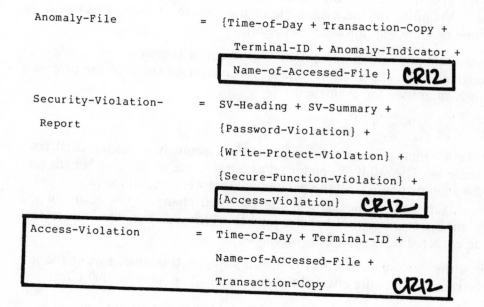

Anomaly-File   = {Time-of-Day + Transaction-Copy +

                               Terminal-ID + Anomaly-Indicator +

                               Name-of-Accessed-File } CR12

Security-Violation-   = SV-Heading + SV-Summary +

  Report   {Password-Violation} +

                              {Write-Protect-Violation} +

                              {Secure-Function-Violation} +

                              {Access-Violation} CR12

Access-Violation   = Time-of-Day + Terminal-ID +

                              Name-of-Accessed-File +

                              Transaction-Copy CR12

Figure 96

3. incremental Data Dictionary to modify the Data Dictionary in the Structured Specification

4. process descriptions to replace those affected in the Structured Specification

24.3 Specification maintenance procedures

Having said this much, I acknowledge that the procedures for specification maintenance may seem obvious. Just for the record, however, I note the following sets of procedures for formulating, evaluating, incorporating, and implementing changes.

24.3.1 Describing a Proposed Change

Each time a change to the Structured Specification is proposed, the analyst prepares a Specification Increment Document. The SID is circulated back and forth between the analyst and all affected parties until it is judged to be an adequate description of the proposed change. During this period, the emphasis is on developing the clearest possible statement of the change and its interface to the existing system model, rather than on trying to decide whether or not it is desirable. Depending on the magnitude of the change, it may be advisable to conduct one or more walkthroughs of the SID.

Once the Specification Increment Document is complete and considered by the originator and the analyst to be a proper representation of the proposed change, it is submitted for evaluation.

24.3.2 Quantifying Impact

The evaluation of each change request is essentially an incremental cost-benefit analysis. The study is conducted as a miniature of the cost-benefit portion of the analysis phase. All the parameters that were quantified in that phase must be reexamined in the light of the proposed change. The result of the study is an incremental cost-benefit report, which declares the impact that should be expected if the change is accepted.

Since the development effort is in full swing at this time, part of the impact to be considered is the effect the change is likely to have on that effort:

- manpower and elapsed time to be added to the current budget and schedule

- saved manpower and elapsed time (due to a simplifying change)

- wasted resources already expended on features that would be removed or supplanted by the change or some combination of these items. To facilitate actually making the change if it is accepted, I encourage you to present this information in the form of an implementation plan for the change, with cost accounting data keyed to it. It might look something like this:

IMPLEMENTATION PLAN FOR CHANGE REQUEST 31

1. Throw away module xxxx (already coded). Wasted effort: 6 work-weeks; saved test and integrate time: 16 work-weeks.

2. Delete module yyyy (still to be coded). Saved effort: 16 work-weeks.

3. Revise module zzzzz to include suchandsuch feature. Added effort: 4 work-days.

All of the associated cost-benefit factors should be weighed against the change and used as input to the decision-making process. If the change is then accepted, the SID is used to update the Structured Specification.

24.3.3 Incorporating the Change into the Structured Specification

This process should be trivial. The SID is constructed so as to be easily incorporated into the Structured Specification. By the time it is used to make a formal update to the specification, each of the parties involved in formulating and evaluating the change will have gone through the steps of incorporating it into their private copies of the Structured Specification (on a temporary basis) in order to consider the change in context. The formal update works the same way. Guided by the indication of affected components (this information is written on the Change Request Form), portions of the incremental model replace portions of the existing model. The result is a revised Structured Specification that describes the system with the accepted change in place.

24.3.4 Implementing the Change

Now the implementation plan, generated as part of the evaluation process, is put into effect.

Depending on the change itself, there is a certain bottom-line minimum effort associated with implementing it. That effort is essentially what it would have taken to implement the portion of the system specified by the change — if it had been included in the original Structured Specification — plus the work that must be thrown away or undone as a result of accepting the change. I refer to this bottom-line minimum effort as the *inherent cost of the change*. Our goal is to incur little or no additional expense and effort beyond the inherent cost of the change. Such additional expense and effort would have to be considered a reflection on the flexibility of our evolving system. The difference between the actual cost to implement a change and the inherent cost of the change is effectively the *cost due to inflexibility*. We can never determine this cost exactly since the concept of inherent cost of a change is rather vague. But clearly we want the cost due to inflexibility to be as small as possible.

Put yourself in the position of a project manager asked to implement a given change (specified in SID format). What are the factors that are likely to control the inflexibility component of the total cost of implementing the change? Without going into all the factors, I think you will agree that the design is the one that will have the greatest effect on expected inflexibility cost. And, in particular, the extent to which the design takes it shape from the system model is going to determine the ease of implementing changes to that model. The most fortuitous situation is one in which there are clear correlations between components of the model (the primitives and data flows, for instance) and components of the design structure (modules and interfaces). That makes it simplest to determine the affected domain of any SID, to evaluate the impact of the change, and to implement the change.

The idea of maintaining correlations between the system model and the design — of deriving the design from the model — is fundamental to Structured Design. And one of the major goals of Structured Design is minimization of inflexibility costs.

This provides a good bridge to our next subject, Structured Design. But first, I am going to sneak in one nasty comment about the concept of Change Control.

24.4 The myth of Change Control

The section heading was my nasty comment.

As soon as the analysis phase ends, someone is sure to declare that the specification is frozen, and that any change which is proposed from now on must be routed though the intimidating and bureaucratic institution known as Change Control. The implication of this is that everybody had his opportunity to shape the new system during analysis, and now that opportunity is past. New and modified requirements will be greeted with derision and scorn. Some will be accepted, and some will be rejected with terse comments. The august body that gets to generate derision, scorn, acceptances, rejections, and terse comments is the Change Control Committee (or some such grandiose name).

If you don't have some mechanism for controlling change, of course, you can't have a frozen specification. It is as simple as that; the two go hand in hand. When I say that Change Control is a myth, I definitely mean to say as well that the frozen specification is a myth. The fact is that change is inevitable. The idea of "freezing the specification" is about as realistic as shooting at a moving target with your eyes closed. You can do it, but you will miss the target because of it. The more successful you are in freezing the specification, the more likely your project is to deliver a system that does not fill the true needs of your user. You and your user can discuss whose fault that was when you meet on the unemployment line.

It is a fallacy to talk about controlling change. The best you can hope for is to keep track of change. If there is some sort of a Change Control Committee working with your project, I suggest that it ought to be renamed the Change Documentation Committee and its charter modified accordingly.

25 TRANSITION INTO THE DESIGN PHASE

The purpose of this chapter is twofold: to provide a glimpse at the structured methods used to conduct the design phase, and to show how the end of analysis dovetails with the early part of the design effort.

This will not be a complete description of the subject of design, or even of those various strategies and methods which are loosely termed Structured Design. When you're done with this chapter you won't know how to do a Structured Design, unless you knew that already. But you will have a clear idea of what Structured Design is all about, what its tools are, and what relationship it bears to Structured Analysis. In addition, I hope to give you enough of a working understanding of the concepts to allow you to pursue the matter further, should it interest you. With that end in mind, I have included in the Bibliography references to five key works on modern design. You will find them under the names Dijkstra, Jackson, Myers, Orr, and Yourdon and Constantine.

In considering the process of design, we need to look into the following:

1. *Goals for design:* What are the economics that ought to guide us in deciding what constitutes a "good" design? What characteristics would such a design have? What is a good design worth? How much should we be willing to give up for it?

2. *Design strategies:* How is the overall design structure determined? How does it relate to the Structured Specification? What are the alternatives?

3. *Design tools:* How do we go about constructing a design? How do we document it? How do we measure its quality? How do we improve it?

There is a section below on each of these topics. Let's begin by considering design goals.

25.1 Goals for design

Would it surprise you to find that your organization spends more money modifying old systems than building new ones? It shouldn't. Many organizations do. Figures from Barry Boehm's study, "Software Engineering" (cited in your Bibliography), indicate that more than 40 percent of software dollars spent each year in the United States is spent on maintenance, on the modification and continued debugging of production systems. Barely half of the total lifetime software cost has been expended at the time that any system goes into production. Worse still, those figures are made disproportionately *optimistic* by the fact that so many systems are abandoned after a productive life of only one or two years. (If you abandon a system immediately upon completing it, the maintenance cost is zero. But that doesn't imply that you have solved the maintenance dilemma.) As the average life of a system increases slowly toward six years, the average percentage of the lifetime software cost devoted to maintenance approaches 60 percent! (Figures again from Barry Boehm.)

These facts should not be new to you — they have appeared with increasing frequency in the industry press — but I hope that you feel, as I do, that the dismal situation they reveal is totally unacceptable. It doesn't make sense to spend four dollars to build, and then six dollars to fix up what you have built. As long as we continue that pattern, there is something grievously wrong with our approach to development. Because so much of our software resources are used up during the maintenance period, it is only reasonable that our first design goal should be to maximize the ease of modifying the system. Relatively small percentages of the lifetime cost of the system are spent on analysis, design, and coding. Because there is so much to be gained by reducing the cost of maintenance, it makes good sense to dedicate extra development effort if that will help. In particular, if an additional 10 percent expenditure on design and coding could reduce maintenance costs by 10 percent, the trade-off would be overwhelmingly favorable.

If it seems intrinsically wrong to you that half the software money should be spent on fixing up delivered systems, perhaps you will consider this even worse: Half the cost of delivering a system is spent in testing and debugging. That means that for every dollar spent in true development (designing and coding), three dollars are spent in revision, either before or after delivery. Again, the high cost of debugging and testing may not be a surprise, but it is a sign that something is very wrong with our methods. Therefore, our second design goal should be this: The systems we design ought to be easier to test and to prove out than systems have been in the past. We should be willing to spend some extra time in design if it can make debugging significantly easier.

Note that both maintenance and debugging costs are people-intensive. So, as the cost of manpower continues to grow, these factors become more and more important. At the same time, machine costs are decreasing significantly. If there is any possibility to trade off people-intensive effort against machine resources (CPU time, disk and core space, seek time), we are likely to benefit from it.

Along with the two major goals of maintainability and ease of testing, I shall add some minor ones: It would be convenient if the documentation of design was a natural by-product of the design process (thus saving us the drudgery of writing a design opus); it would be convenient if the design structure made it easy to isolate the effect of a given change; it would be convenient if the design allowed subsequent development efforts — coding and debugging — to proceed with little need for communication among the implementors; it would be convenient if the design methodology caused convergence so that two different designers working on the same problem tended to come up with the same or similar solutions; it would be convenient if the design methodology caused a large degree of partitioning, since that would allow more flexible work allocation during implementation; finally, it would be convenient if the design method resulted in a smooth progression from abstract to detailed (i.e., displayed the top-down characteristic), so that the more senior personnel could work on the architecture of the design, while the junior designers worked on the details.

25.1.1 The Maintainable System

What design features contribute to ease of maintenance and modification? In his early work at Hughes Aircraft and IBM, Larry Constantine tried to derive an empirical answer to this question by looking at large numbers of systems for which cost accounting data had been collected, and correlating observed low cost of maintenance with characteristics of design. I list here some of his observations, along with my own intuitive explanation for each one.

Ease of maintenance was strongly related to the following design features:

1. *Small module size:* The smaller the modules, the more likely it is that the impact of a given change (or fix) can be effectively isolated. The number of modules affected by the change will increase as module size declines, but the total amount of code that must be considered is reduced.

2. *Modular independence:* The less the inside of one module depends upon anything that is inside another module, the easier it will be to maintain. The extent of modular independence determines the feasibility of working on modules (reading them, debugging them, modifying them) one at a time. A module's independence from its neighbors is decreased by control connections (branches and calls to and from other modules), shared data, shared files, shared devices, and common interfaces to the operating system.

3. *Black-box characteristic:* The black-box characteristic applies to any system, program, or module that can be viewed in terms of its inputs and outputs alone, without worrying about the details of how it processes them. It has a strong effect on readability. The maintenance programmer ought to be able to read through code, guided by his own requirements. But some systems are written so that they must be read in one particular way, regardless of the particular interests of the reader. A sys-

tem that is particularly readable is one that can be viewed as a set of nested black boxes; i.e., there is one top-level module (the main program) which contains both the system entry point and the system exit. It is effectively the black box that contains the whole system. The module itself consists of calls to other modules, each of which can be viewed as a smaller black box; those modules consist of calls to still smaller black boxes, and so on. The maintenance programmer reads the top-level module to get the big picture, and then *chooses* which one to read next. When he reads that module, he gets the big picture of the function allocated to that particular module, and then *chooses* which of its subordinates to look into next. Modular systems which exhibit the nested black-box characteristic throughout are particularly readable, hence easier to modify and debug.

4.   *Conceptual modeling:* The structure of any system takes its shape from something. That something is the model for the system. A team of compiler writers turned loose on a cost accounting system might make it look internally like a compiler (it would have stacks and reverse Polish notation). In that case, the model for the system would be their particular limited field of experience. For a system to be maintainable, it should be modeled on something conceptually understandable and familiar to the maintenance engineer. That means it has to be modeled on the nature of the problem it is intended to solve. This is rather a negative statement — it suggests a host of very familiar system models that should never be used. In particular, systems should not be modeled on: the machine, the operating system, the organization (if you can see the makeup of the different implementation teams reflected in the design, it is a bad one), the budget (if you can see the various project phases reflected in the design, it is a bad one), or any general-purpose model intended to be applied to all systems regardless of what they do. In this last category I include familiar models like Input-Process-Output and Read-Edit-Update-Print, models that have accounted for more than their share of bad designs over the years.

5.   *Isolation of detail:* There ought to be reasonable separation between those parts of the system which reflect its underlying philosophy, and those parts which reflect mere details. The reason this is important is that the details are much more likely to change than the philosophy. It is unacceptable for the two to be so intermixed that a change to any detail may cause a drastic revision of the system's fundamental nature.

For a more complete treatment of this material, I recommend Yourdon and Constantine's *Structured Design*.

25.1.2 The Buildable System

The same factors that control maintainability have a strong effect on how simple or complicated it is to test the system. The relative importance of these factors is slightly different, though, when viewed from the perspective of the implementor rather than the maintainer. During testing, the independence of the modules and the black-box characteristic are particularly important. The possibility of isolating detail at the lowest levels (with a smooth progression from abstract to detailed) is essential if a top-down implementation scheme is used. More on top-down implementation at the end of this chapter.

25.1.3 Efficiency

The topic of efficiency is included here not because I want to reemphasize its importance as a design goal, but because I believe that it has always been overemphasized. Designers and programmers have accepted that the system and every component of it had to be highly efficient as an unspoken objective. That means coders have slaved away to tighten up modules that only run during summer Sunday afternoons in alternate leap years. Good practice has dictated that all code should be as fast as possible and use up the minimum amount of space.

Our preoccupation with efficiency is a leftover 1950's design goal. It dates from the days when computers cost two orders of magnitude more than they do today. There are still applications where efficiency is important, but it is never as important as it used to be. It is almost never as important as the cost of maintenance and testing. Even Bell Laboratories, the builders of switching systems that are replicated thousands of times, have come to the conclusion that they can afford to sacrifice a lot of efficiency to gain a small saving in maintenance. In systems where efficiency is essential, usually only a small percentage of the modules have to be highly efficient. Nobody cares how quickly the others run or how much space they take up, because they are used so seldom.

My point in trying to relegate efficiency to the status of a secondary design goal is this: Flexibility and efficiency are opposite sides of the same coin. The more efficient a system is, the harder it will be to modify. (That 14-instruction loop that the coders have been tuning up for years is a real maintenance disaster; if it has to be changed, your entire maintenance team is liable to quit or commit suicide.) This strong negative correlation between efficiency and flexibility applies not only to EDP systems, but to systems of all kinds.

It applies, for instance, to genetic systems. An organism (genetic system) that is highly efficient in its own environment is relatively unadaptable should the environment change. The giraffe is a perfect example. It has evolved into the ultimate life form for an environment where the food source is high in the branches of trees. Any place else, it is a marginal survivor. The geneticist R.A. Fisher expressed this idea in what has come to be called Fisher's Fundamental Theorem: The more highly adapted an organism is to its present environment, the less adaptable it will be to any other.[1] In our own terms, the more efficient a computer system is, the harder it will be to modify.

The price paid for ease of maintenance is probably going to be paid in terms of efficiency. You should expect a system designed for flexibility to be slower and to use up more space. Empirical figures collected by Yourdon inc. indicate that the efficiency penalty due to the use of Structured Design is on the order of 10 percent in each of these categories.

25.2 Structured Design

Structured Design is a specific approach to the design process, an approach that results in small, independent, black-box modules, arranged in a hierarchy that is a conceptual model of the business area, organized in a top-down fashion with the details isolated at the bottom. In other words, it is a strategy for designing maintainable, provable systems.

The major concepts of Structured Design are usually attributed to Larry Constantine, whose work on the subject dates back to the late 1960's. Larry is the coauthor, with Ed Yourdon, of an important text on the subject and, with Stevens and Myers, of the *IBM Systems Journal* paper that brought the term Structured Design to public attention. Both of these works are referenced in the Bibliography.

Before we plunge into the subject, here are a few relevant definitions.

- *A design strategy* is a set of methods for deriving, evaluating, refining, and documenting a design.

- *A top-down design* is a particular kind of design rather than a strategy. A design is said to be top-down if it is made up of a hierarchy of modules, each one a single-entry, single-exit subroutine or the equivalent.

- *Structured Design* is a strategy for producing a highly maintainable, easily tested top-down design. The derivative techniques of Structured Design are transform analysis and transaction analysis. Its evaluative and refinement techniques are coupling, cohesion, and packaging, together with some miscellaneous design heuristics. The documentation tool for Structured Design is the Structure Chart. Each of these topics is discussed below.

---

[1] Ronald A. Fisher, *The Genetical Theory of Natural Selection*, Dover Publications, 1958.

25.2.1 The Basic Idea

Any computer system or program has both procedural and hierarchical characteristics. Its procedural characteristics define the *order* that governs processing, the idea that x happens first, then y, and so on. Its hierarchical characteristics define the *rank* of the various components of the design, the idea that this module is the boss, this one is a level-two subroutine, while these are absolute bottom-level subroutines that have no subordinates. We shall need to determine both the procedural and hierarchical nature of our design before it can be considered complete. But where shall we begin the design process?

Classical design begins with consideration of the procedural characteristics. The designer's thinking as he begins to work out the shape of a system (a system to control a space vehicle, for instance) might go something like this:

Q: What happens first?

A: We initialize.

Q: What do we initialize?

A: Who knows? But when we figure out what to initialize, this is where we'll do it.

Q: What happens next?

A: I guess we better get some input.

Q: What kind of input?

A: Well, whatever turns up.

Q: And then?

A: Then we'll edit it.

Q: With respect to?

A: Oh, files and tables and stuff.

When you do a procedural design, you use a procedural design tool — such as a flowchart — to document your concept. Fig. 97 shows what you might have for your initial design sketch of the space vehicle guidance system if you went about it in this fashion.

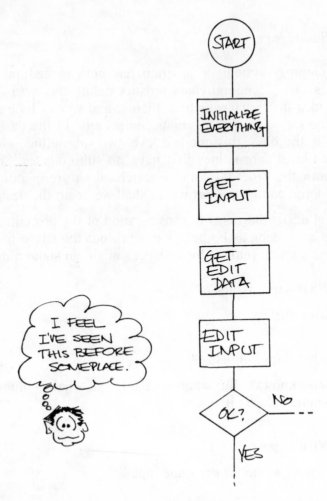

Figure 97

By starting with what happens first, you necessarily begin with the details. Initialization is a detail; it does not have anything to do with the nature of the system. When you start out with the details, you build a bottom-up design.

Structured Design begins with the hierarchical characteristics. It encourages you to deal with questions in the order of their importance rather than in the order in which the computer will face them. The same designer considering the same system as above, but this time using Structured Design, might approach it like this:

Q: What is the nature of the system?

A: It guides a space vehicle.

Q: What are the most important aspects of this work?

A: Obeying control commands from the pilot, reading positional data from the sensors, issuing orders to direct and fire the rockets, and monitoring all the internal checkpoints.

Just as a flowchart serves to document a procedural view, a *Structure Chart* serves to document the hierarchical view of a system. Fig. 98 shows a Structure Chart of the initial design sketch of the space vehicle guidance system; Fig. 99 shows this same Structure Chart after the next iteration of the design. As you can see, a second level has been added to show more detail. If you are not familiar with the mechanics of Structure Charts, the next section will explain them to you. But even without knowing precisely how to read Fig. 98, I think you can see that it addresses concepts much more vital to the application than the flowchart does. It somehow *looks like* a space vehicle guidance system, while the flowchart just looks like a flowchart.

The basic idea of Structured Design is this: The design should take its shape from a hierarchical view of the application, not a procedural view. The top level shows the most important division of work; lower levels further subdivide the work allocated to each of their managers. The underlying philosophy of the system appears at the top, and the details at the bottom.

25.2.2 Notational Conventions for Structured Design

A Structure Chart is a graphic tool for representing hierarchy. The Structure Charts in Figs. 98 and 99 are drawn according to the Constantine convention, but that is not the only possibility. This kind of Structure Chart has three basic elements:

- *The module,* represented by a rectangular box with a name on the inside. A module is a named, bounded, contiguous set of statements.

- *The connection,* represented by a vector joining two modules. A connection is any reference from one module to something defined in another module. Usually it means that one module has called the other.

- *The couple,* represented by a short arrow with a circular tail. A couple is a data item that moves from one module to another.

A connection may be annotated with one or more couples showing that data items are moved as part of the connection. Fig. 100 portrays several annotated connections and their meanings.

Now let's go back and interpret the Structure Chart of Fig. 98. It shows seven modules; the one called Guide Space Vehicle is the president since it manages (has connections to) all the others. All the connections are normal; the president module calls each of the subordinates, and they each guarantee to return. The data passed between the president and its subordinates is shown by the couples. (There should be an entry in the design phase Data Dictionary for each couple to define its composition.) By convention, all shared data is declared on the Structure Chart using couples. That means there is no shared

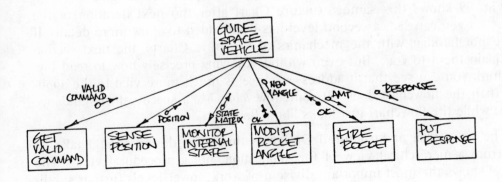

Figure 98

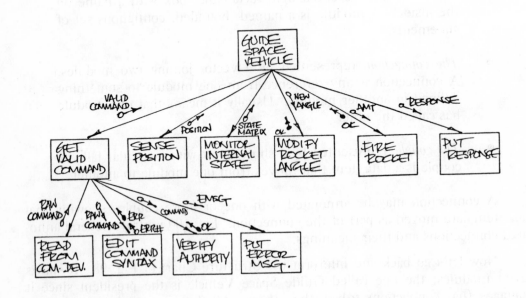

Figure 99

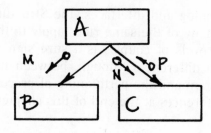

Two normal connections:

1. A calls B, passing M downward.
2. A calls C, passing N downward and receiving P back.

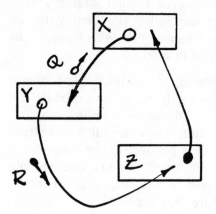

Three pathological connections:

1. X reads Y's private variable, Q.
2. Y over-writes Z's private flag, R.
3. Z transfers control into X.

Figure 100

data in addition to what is shown. (Modify Rocket Direction does not peek at the internal data of Fire Rockets — if it did, a pathological data connection would have to be shown.)

Fig. 99 shows an additional level of detail under the vice-presidential module called Get Valid Command. As you can see, it subcontracts some of its function to the third-level subroutines that read the command, edit its syntax, check the authority of the sender, and put out an error message if one is required. It calls each of these modules as needed. The set of connections states that there is at least one call inside Get Valid Command to each of its subordinates. It doesn't say anything about how many more calls there may be, how often they are used, whether they are used, or in what order they are used. All of that procedural information is omitted from a Structure Chart.

A Structure Chart shows the partitioning of a system into modules, the hierarchy into which the modules are arranged, and the interfaces among modules. It doesn't tell anything about the decision structure of the system or the order in which the various functions are performed.

In the sense that it emphasizes partitioning and interfaces, the Structure Chart is rather like a Data Flow Diagram. Many of the same rules apply to the two; for instance, a Structure Chart with crowds of couples is a sure sign of poor partitioning. There are some important differences, though, between the partitioning of function on a Data Flow Diagram and the partitioning of process on a Structure Chart. We'll look at those differences at the end of this chapter.

25.2.3 Coupling

The topic of coupling is central to the Structured Design strategy. It is through coupling evaluation that the designer can judge the quality of his design and see which areas need improvement. When he makes a change, coupling helps him to determine whether it has helped or hurt.

Coupling is a measure of the interdependence of modules. The higher the coupling, the more likely it is that changes to the inside of one module will affect the proper functioning of another module. Obviously coupling is something we would like to minimize. There is no way to make modules in a structure absolutely independent of one another, but it is possible to come up with a structure that has so little coupling that you can *usually* modify one module without disrupting others. That is a structure that is *usually* easy to maintain. Since we are unlikely to come up with a structure that is always easy to maintain (i.e., impervious to any change), we shall have to be content to be able to make most changes with minimal difficulty.

Coupling also has a strong effect on the readability of both the design and the code of each module. If coupling is minimized, you can read through a given module without having to look inside any other module in order to complete your understanding. The data is all declared right there, either as private data or as passed parameters. There is no untoward branching into the module in odd places from outside, nor does the module itself branch into the interior of others. No other module has undeclared access to what is apparently this module's own data. Its interfaces to the rest of the world are minimized and stated explicitly.

How much does the quality of a given design depend on coupling? Glenford Myers, in his book *Reliable Software Through Composite Design*, puts it this way:

> "A quantitative measure of the independence of two modules could be developed based on the coupling between them."

In the dry and technical world of EDP systems design, that statement is a blockbuster. Myers is saying that if you drew a Structure Chart of your system and declared each and every item of data that moved between two modules, he could tell you how much overall modular independence the structure had. He could express that with a matrix of probabilities; such a matrix would tell you the probability of a change in module n causing a bug to crop up in module m.

And he would tell you what factors affected the probabilities so that you could know what to do to reduce them. Each time you altered your design concept, you could apply his "scoring algorithm" to see if you had improved modular independence.

Take a look at Myers' work if you are interested in the details of how the quantitative evaluation would work and what its restrictions are. But briefly, these are the factors that Myers and others have found to affect coupling:

1. *The type of connection* can have such a strong influence on coupling between modules that certain kinds of connections are virtually forbidden in Structured Design. For instance, use of the GOTO to transfer control between modules couples the modules together almost hopelessly; it makes a mockery of any attempt to deal with modules one at a time, since you cannot even tell under what circumstances any piece of code is entered. Data connections (references from one module to another module's data) are also costly. They introduce invisible coupling (coupling not expressly declared in the code of the module that currently has the data), and that reduces the readability of all modules. Undeclared connections such as GOTO's and data connections are termed "pathological" for their effect on coupling.

2. *The type of data passed* between two modules affects their interdependence. Computational items (used for calculating and indexing) have less effect than switching items (used as a basis of a decision). This has been judged important enough to require the addition of a notational distinction between computational couples and switching couples. This is illustrated in Fig. 99 by the switching couple "ERR" and the computational couple "ERR #" passed back from the syntax editing module.

3. *The amount of data passed* between two modules is also important. The number of couples and the size of each couple figure into the calculation of interdependence.

4. *The direction of some couples* is also relevant. For instance, downward passing switches have a stronger effect on linking modules together than upward passing switches. This is because downward passing switches tend to ruin the integrity of the receiving module, by driving it from above to do things whose significance it cannot fully understand.

There are other factors as well that affect coupling. My intention here is not to list them all and show you how to look at them to determine modular independence, but only to indicate that the quality of a design is strongly influenced by the amount and nature of data that moves among modules.

25.2.4 Cohesion

Cohesion is a good quality exhibited by some design structures. Before I define it, look at Fig. 101, an alternate Structure Chart for the space vehicle guidance system we considered earlier. Fig. 101 is an abominable design. It is proof positive that one can design poorly even using a Structure Chart. ("Plowin' ain't potatoes.") What the design of Fig. 101 lacks is cohesion. Every module on the figure is weakly cohesive.

Fig. 99, on the other hand, is made up of strongly cohesive modules. By comparing the two figures, you can probably see exactly what cohesion is. It has to do with the integrity or "strength" of each module. The more valid a module's reason for existing as a module, the more cohesive it is.

Cohesion is a measure of the strength of association of the elements inside a module. A highly cohesive module is a collection of statements and data items that should be treated as a whole because they are so closely related. Any attempt to divide them up would only result in increased coupling and decreased readability.

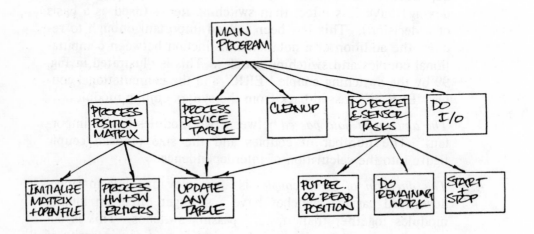

Figure 101

Some designers have tried to formulate a whole range of levels of cohesion, varying from the worst to the best. For the purposes of this book, however, I propose to use only two levels: acceptable and unacceptable. Modules are of acceptable cohesion when they do one and only one allocated task, or they do several related tasks that are grouped together because they are strongly bound by use of the same data items. Modules have unacceptable cohesion when they perform unrelated tasks, bound together only by weak rationalizations, time dependencies (things that happen at the same time), or order dependencies (things that happen one after the other).

There are two standards used to judge a module's cohesion: its name and its coupling. If you look back to the weakly cohesive modules of Fig. 101, you will see that none of the names is very helpful. Many of them use weak verbs ("process," for instance) or multiple unrelated verbs. Several, such as Update Any Table and Process HW and SW Errors, have a distinctly general-purpose sound to them. The module that takes care of hardware and software errors presumably does everything from putting out warnings and error messages, to blowing up the ship when the enemy is near.

If you can't think of a meaningful name for a module ("What do you call this module?" "Arthur."), then it is almost certainly unacceptable. If the best name you can come up with is wishy-washy or involves multiple verbs and objects, the module is of unacceptable cohesion. If you can come up with a strong name made up of one transitive verb and a single object, the module is likely to be strongly cohesive.

Cohesion does not only apply to the bottom-level modules. The whole structure should be strongly cohesive. For your naming test to have some meaning in the upper levels, you have to adopt the convention of naming from above. This means that the name given to each module is not only a description of what goes on inside the module, but also of its entire allocated responsibility (just as though it had no subordinates). The name of the module is the name of the entire tree. The name of the top module, by this convention, is the name of the whole system.

There is often a strong relationship between coupling and cohesion. In fact, I have not even tried to fill in the coupling for Fig. 101, because such flagrant examples of poor cohesion are invariably intimately coupled (nearly every module has connections to every other module). Beyond the simple correlation between overcoupling and poor cohesion, there are some significant coupling patterns. These patterns and their implications on the cohesion of the modules are shown in Fig. 102. Modules coupled with a single input couple and/or a single output couple are likely to be of the strongest cohesion. Modules with downward passing switches to tell them what to do are likely to be of poor cohesion. Modules that require pathological data communication are likely to be of poor cohesion.

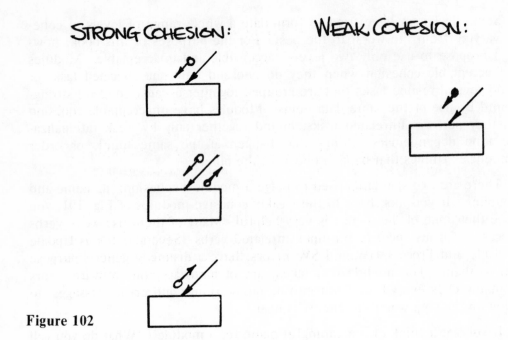

Figure 102

The downward passing switch is probably the simplest test of poor cohesion. Look again at Fig. 101 and try to fill in some of the required downward passing switches. Each module except the top needs at least one. Process Position Matrix, just to take one example, needs switches to tell it whether to add or delete or extract information to/from the matrix, whether to initialize the matrix, and whether to expect passed parameters.

25.2.5 Packaging

Packaging refers to the act of shaping the design to accommodate the physical environment: the machine, operating system, coding language, core and disk limitations, time restrictions. The different slant that Structured Design gives us on packaging is not so much how to go about it, but *when* to go about it. The key Structured Design packaging principle is this:

> All packaging considerations are put off until the rest of the design is complete. An environment-independent design is first constructed to maximize cohesion and minimize coupling. Packaging requirements are then imposed on this ideal design in such a way as to introduce the minimum degradation of its quality.

This is quite the opposite of the more traditional approach in which packaging considerations were allowed to shape the whole system:

Q: How shall we begin?

A: By keeping in mind that everything has to be coded *very tightly* because this system has to run in the minimum amount of core.

Q: Is that to be the guiding concept?

A: That and the very curious nature of the drum we're using for overlays. It has a sector size of 913 words. We definitely want our modules to be, as nearly as possible, 913 words each.

Q: Yes, but what is our system to be modeled upon?

A: Well, we want it to be written in COBOL, that is certainly one consideration. I understand that COBOL is very slow to do a PERFORM VARYING. So our design ought to be one that avoids those.

Q: Yes, but . . .

A: And don't forget OS!!! OS is a terrible pig if it has to use the linkage section at all. And let's, by all means, have good big compilations so all our listings stay together. And since disk space is so limited, we'd better . . .

The problem when packaging shapes the system is that its effect cannot be controlled. If you set out to make the system enormously efficient, for example, it may end up efficient throughout, and correspondingly impossible to modify. If you package after the fact, then you can tighten up only those parts of the design where efficiency really matters. The rest can be left in its pristine state and will therefore be as flexible and provable as possible.

When should changes due to packaging be introduced into the design? As late as possible. At the risk of sounding like an extremist, I would encourage you to delay all packaging until your system is up and running. That way, you get the maximum benefit from your unmodified design during the crucial testing process. Then, after the system is demonstrably correct, you can begin to tune it. If it is efficiency it lacks, you can use real tuning aids (hardware monitors and software resource analysis packages) to guide you. I quote Ed Yourdon on this subject:

"It is easier to make a working system efficient than to make an efficient system work."

By the way, if you do defer tuning until the very end, I predict you will find yourself making different modifications from those you would have expected — the inefficiencies in a system are almost never what your intuition would have had you believe. So not listening to your efficiency intuition at the very beginning is no great loss.

The following is my list of packaging considerations, all of which should be deferred as long as possible:

- efficiency (CPU efficiency, use of core and disk, compile time and link time efficiency, efficient use of operating system features, and so forth)

- coding and language concerns

- overlay structure

- linkage conventions

- grouping of modules into programs

- passing parameters as opposed to pointers

- hardware dependencies

During design, try not to let any of these considerations influence you at all. Assume efficiency doesn't matter, that you have a picopicosecond machine with a trillion bytes of memory. Compile your modules separately, even the tiny ones, and use a calling sequence to pass parameters — that way, the compiler and linkage editor will enforce your coupling rules for you. Assume that there is no COMMON, at least while you are designing (that means there is no possibility of sneaking in an undeclared pathology). Ignore the idiosyncrasies of your hardware as long as possible: If your machine has a tendency to drop bit 13 of the linking register, don't let that fact determine design structure — you may get a different machine, and wouldn't you look silly then?

25.2.6 Structure Charts and Data Flow Diagrams

Drawing a Structure Chart is an exercise in hierarchical partitioning. It involves analysis of the interfaces as a feedback mechanism to determine the quality of partitioning. If that sounds familiar to you, it is because the same observations could be and have been applied to the exercise of drawing a set of leveled Data Flow Diagrams. There is obviously a strong correlation between Structure Charts and Data Flow Diagrams, and a correspondence as well between the ways the two are used.

A DFD is a statement of requirement. It declares *what* has to be accomplished. A Structure Chart is a statement of design. It declares *how* the requirement shall be met. The relationship between the two reflects the relationship between intent and method. This relationship was elegantly expressed by a late 19th century design philosopher and architect, Louis Sullivan:

"Form ever follows function." [2]

---

[2]Louis Sullivan, "The Tall Office Building Artistically Considered," *Lippincott's Magazine,* March 1896.

It was upon this premise that the Bauhaus school of design was founded in 1919, and the idea has affected design thinking ever since.

The concept that form follows function means that a design must somehow be *derived* from its requirement. A Bauhaus purist would say that there is only one right design for any given requirement, and that the design process consists of finding, not inventing it. A complete and accurate statement of the requirement *is* the design.

That implies that there should be some way to derive a computer system design from the associated specification, to derive a Structure Chart from the associated Data Flow Diagram. In fact, there are two ways: transform analysis and transaction analysis. These are the two disciplines of Structured Design that guide you in your initial design decisions. They tell you how to begin the design process.

The two disciplines deal with two different kinds of Data Flow Diagrams, or portions thereof. Transform analysis applies to applications that are transforms — that is, applications that have clearly identified input streams, central processing, and output streams. A transform is represented in Data Flow Diagram terms by a linear network. Transaction analysis applies to transaction centers, parts of the application characterized by sudden parallelism of data flow. Fig. 103 presents Data Flow Diagrams in the shape of a typical transform and a typical transaction center.

Transform and transaction analysis are cookbook procedures for deriving a Structure Chart from a Data Flow Diagram. But instead of describing the procedures, I am just going to show you the results. Once you understand the relationships that are fundamental to the two techniques (correlations between the DFD and the Structure Chart), the process of deriving Structure Charts will be evident.

Fig. 104 shows the Data Flow Diagram for a transform and the equivalent derived Structure Chart. The essential relationships are these:

1.   The top of the Structure Chart corresponds to the central transform portion of the Data Flow Diagram (the part that is concerned with neither input nor output).

2.   There is one vice-presidential module for each input stream, one for each output stream, and one for the central transform.

3.   Coupling from each input vice president corresponds to the data flow into the central transform from the associated portion of the Data Flow Diagram. Coupling to each output vice president corresponds to the data flow to the associated portion of the Data Flow Diagram from the central transform.

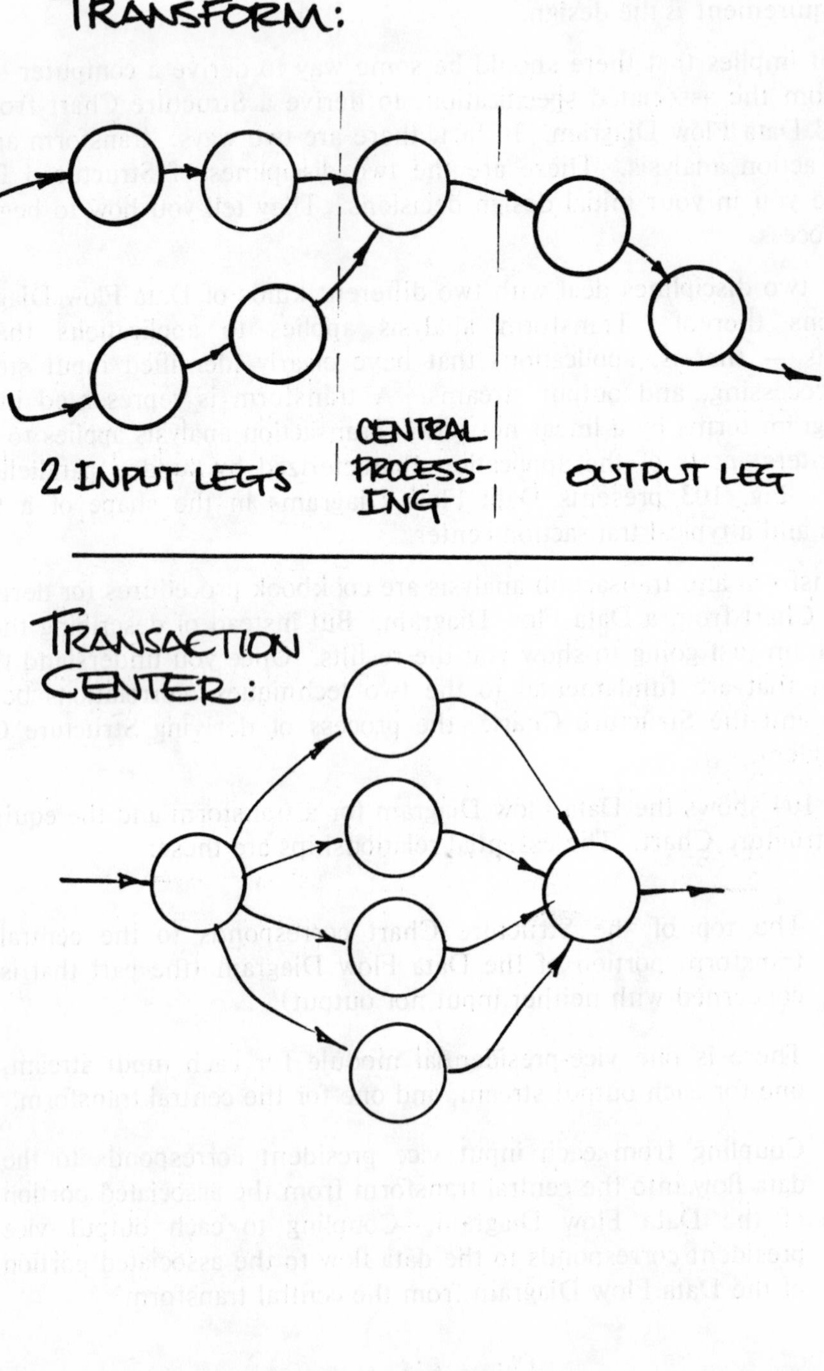

TRANSFORM:

2 INPUT LEGS | CENTRAL PROCESS-ING | OUTPUT LEG

TRANSACTION CENTER:

Figure 103

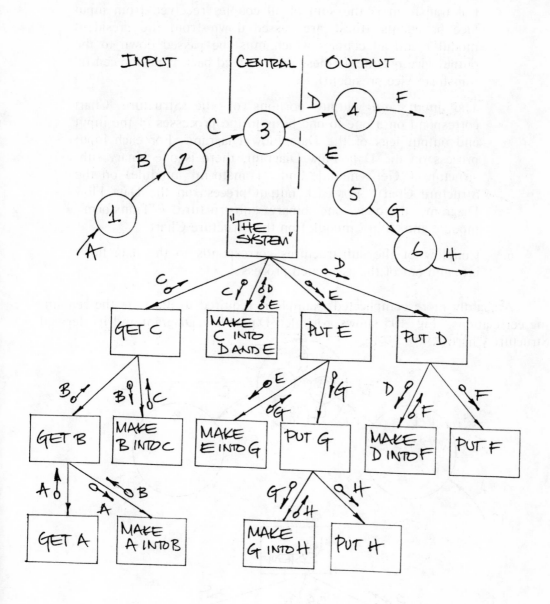

Figure 104

4.   Coupling to and from the vice president in charge of the central transform is the sum of all couples received from input vice presidents (these are passed down from the president module) and all couples which must be passed down to the output vice president (these are received back from the central transform vice president).

5.   The input and output portions of the Structure Chart correspond on a one-to-one basis to the processes of the input and output legs of the Data Flow Diagram. For each input process on the Data Flow Diagram, there is one binary substructure ("Get" module and "Transform" module) on the Structure Chart. For each output process on the Data Flow Diagram, there is one binary substructure ("Transform" module and "Put" module) on the Structure Chart.

6.   Coupling of the substructures corresponds to the data flows into and out of the associated process.

A slightly more complicated example is required to illustrate the remaining correlations. Fig. 105 shows a leveled Data Flow Diagram and its derived Structure Chart.

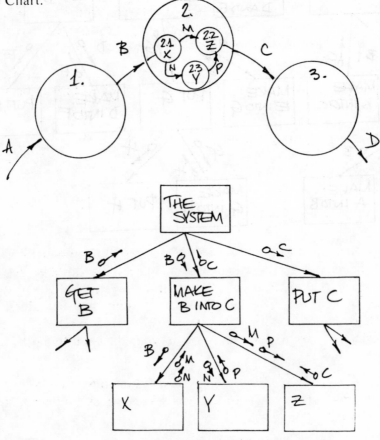

Figure 105

7.   Lower levels of modules beneath a transform module correspond on a one-to-one basis to subordinate processes on the Data Flow Diagram set.

8.   Coupling to and from these subordinates is derived from data flows to and from the associated processes.

Before going on to look at derivation from transaction centers, let's take stock of what we have got. The design process is not over when a Structure Chart has been derived by any simplistic process such as transform or transaction analysis. It has only begun:

Transform and transaction analysis are used to give the Structure Chart its initial shape. The result still requires considerable refinement.

Our Bauhaus purist would be shocked by that heretical statement. It implies that form follows function but not dogmatically — that the design starts with the statement of requirement, and then is altered to cope with certain deficiencies that result from the derivation. This alteration is necessary to give us a good design. The alterations are dictated by considerations of coupling and cohesion, so they tend to improve the flexibility of the structure. One hopes this can be done without deviating too drastically from the initial derived design; that design is the one that is most exactly a model of the requirement, hence the one that is conceptually easiest for all parties to understand.

The reason that the pure model of the requirement is not a perfect design is that some essential factors were not considered at all during the Data Flow Diagramming process:

- flow of control information
- "trivial" error paths
- loops and decisions
- packaging considerations

The emerging Structure Chart must be reconsidered in the light of all of these factors. You have to insert the switching couples (they can't be derived from the DFD's, since the controls were ignored when the DFD's were drawn). Modules may become overcoupled or their cohesion may be ruined by these additions. There is no avoiding the fact that some revisions will be required to refine the derived structure. In fact, most of your design effort will be spent in this refinement.

Now we are ready for transaction analysis. Fig. 106 shows a DFD transaction center and its derived Structure Chart. The correspondences are reasonably straightforward:

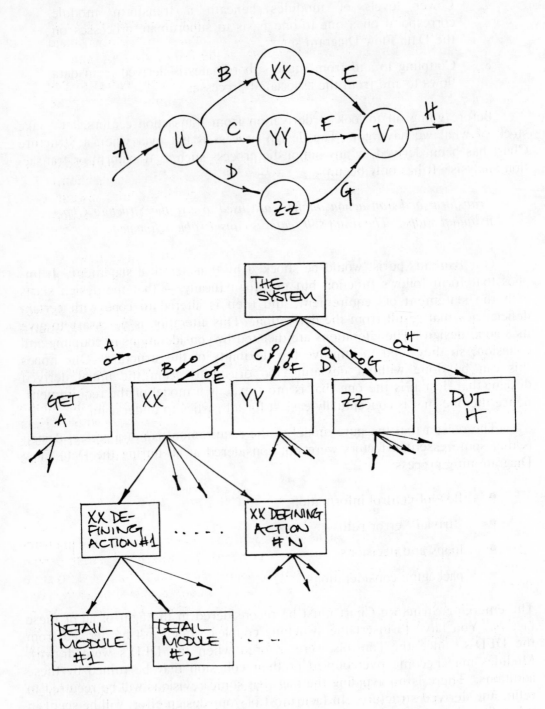

Figure 106

9. The module that corresponds to the transaction center process is itself a transaction center, i.e., it manages one subordinate module for each parallel process on the Data Flow Diagram.

10. Each subordinate module under the transaction center is allocated one of the parallel processes. These subordinates make up the "transaction level."

11. Under each transaction-level module is a set of "action modules," one for each defining action. (The defining actions are the functions that combine to describe the transaction.) Under the action modules is a level of detail modules, usually useful subroutines called by one or more of the action modules.

12. Coupling to the transaction level corresponds to data flows to and from the set of parallel processes. (See detail of Fig. 106.)

As you might have guessed, transaction analysis is an empirically derived approach — so many transaction centers ended up with the three levels (transaction, action, and detail) under them that the idea was formalized.

Of course, it is rare that any application is a pure transform or transaction center. Most of our systems are a bit of each. Fortunately, the deriving techniques can be used on portions of the Data Flow Diagram to determine corresponding portions of the Structure Chart. Fig. 107 shows a complex Data Flow Diagram upon which both techniques have been applied to derive the Structure Chart. Note that when both techniques are used, transform analysis should be used first.

25.2.7 Data-Driven Design

Some modern designers believe it is more profitable to model initial structure upon the structure of the data itself rather than on its flow. These people would have you use the Data Dictionary as principal input to any derivative process. The structure that would result would have an appealing correspondence to the major files and data constructs of the system. For instance, there would be loops in the design structure to correspond to iterations in the data definitions, decisions to correspond to selections, and sets of sequential processes to correspond to data sequences.

I am not a partisan of this approach because I believe that the data structure is more likely to vary (and cause a need for fundamental change to the design structure) than the data flow. However, I must admit that I have made use of the data-driven design philosophy in isolated circumstances and have found it helpful for detailed module design. A basic understanding of it would complement any designer's bag of tricks.

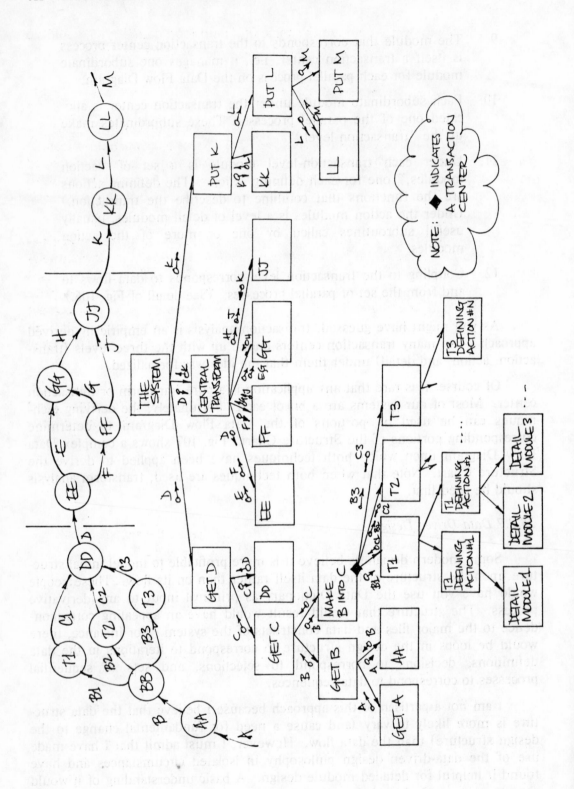

Figure 107

If you would like to pursue the topic of data-driven design, you would do well to read the words of Warnier, Orr, or Jackson (see Bibliography).

25.3 Implementing Structured Designs

You could implement a Structured Design using the familiar phased approach: coding, unit testing, subsystem testing, system and integration testing. But that has one unhappy failing: The most important interfaces aren't tested until the very end. The early part of the implementation is hopelessly involved in the details. That is contrary to the spirit that caused us to revise our analysis and design methods.

When you design in a structured fashion, you have an attractive alternative to phased implementation: top-down implementation. Top-down implementation has you code and unit-test the top module and the second-level modules, and then put them together to test the top of the structure. In place of the third-level modules, you link "stubs," which are skeleton replacement modules that emulate only the simplest functions of the modules being replaced. (A stub might be nothing more than a return with a fixed value passed back.)

Each time you code and unit-test new modules, you fill in another level. You write new stubs for the first level not actually coded, and now you have a new version. Fig. 108 shows two successive top-down implementation versions with their stubs.

The advantages of top-down implementation are substantial:

- The critical interfaces are tested first.

- Early versions, though incomplete, may resemble the real system enough to provide some useful visibility.

- There is no need for costly and complicated test drivers. Stubs are much simpler (but you should expect to have more of them).

- You can have useful overlap of coding, unit testing, and integration testing.

- Machine-time requirements are smoothed out. Morale is improved.

VERSION 1:

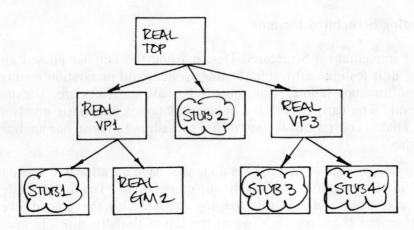

VERSION 2:

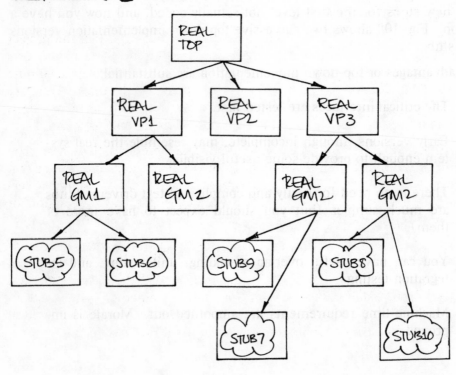

Figure 108

26 ACCEPTANCE TESTING

The purpose of acceptance testing is to avoid unwelcome surprises after the system goes into production. The purpose of Structured Analysis is to avoid unwelcome surprises in acceptance testing. The two are closely related. This chapter will examine that relationship.

During acceptance testing, the loop is finally closed between the requirement and the implementation. Acceptance testing determines whether or not the implementation was on target. Minor corrections may be made during the acceptance test to fine-tune system characteristics that are slightly off target, but acceptance testing is not part of the debugging process. It is more akin to a final exam in a university course, an exam that results in a pass or fail grade. The attitude of the examiner at such a time reflects this idea: The time for correction of errors is past; it is now time to accept or reject the result. We should adopt that same attitude.

The result of acceptance testing is binary: it proves either that the project was a success and the system met its target, or that it did not. Acceptance tests that drag on from one iteration to the next, reducing the list of outstanding "exceptions" from 238 to 116 to 59, are not acceptance tests at all. They are indications of a failed project, one for which the development team has been relieved of its implementation duties and the user is left to fend for himself.

If all this sounds terribly cruel toward the implementors, it is not. The acceptance test does not evaluate the work of the implementors, but rather that of the analyst. If the system doesn't pass muster, the problem is almost certainly in the Target Document — either an improperly stated requirement, or one that cannot be met within the restrictions imposed on the project.

In order for projects to aim at appropriate goals, the Target Document must contain *all* the criteria for success. If the analyst is to expect no surprises from the system's performance during acceptance test, then he must be sure that the test itself entails no surprises for the implementor. It is not in anyone's interest to conceal project targets. Tom Gildersleeve, who has written widely on management of development projects, puts it this way:[1]

---

[1]Informal remarks made at Yourdon inc., April 8, 1977.

325

"The acceptance test *is* the specification."

I prefer to look at it the other way around: The specification *is* the acceptance test. Following this line of thought to its logical conclusion leads us to

The Cardinal Rule of Acceptance Testing: Acceptance tests are derived from the Target Document and from the Target Document alone.

No additional criteria may be introduced in the acceptance test, nothing that is not contained in the Target Document. The procedure by which acceptance tests are derived should be well understood by all parties (user, analyst, and implementors). More important, the fact that acceptance tests will be derived from the Target Document must be understood by all parties, particularly the user. Once he has accepted the Target Document, the acceptance test is determined.

How are acceptance tests determined? In particular, how are they derived from the Structured Specification? The relationship between the Structured Specification and the tests will depend on the type of test. Acceptance tests fall into the following five categories:

1. normal path tests
2. exception path tests
3. transient state tests
4. performance tests
5. special tests

The normal path tests are the simplest so we shall begin with those.

26.1 Derivation of normal path tests

Normal path tests are derived from the Data Dictionary. There is one normal path test set (NPTS) for each input data flow. Each NPTS is made up of some number of valid test inputs, each one conforming to the Data Dictionary definition of the data flow.

An example to illustrate test derivations is the following input data flow:

| Credit-Trans | = | Account-No. + (Customer-Name) + |
| | | $\left\lvert\begin{array}{l}\text{Dollar-Amount}\\\text{Credit-Ref-No.}\end{array}\right\rvert$ |

| Account-No. | = | "A" + 6 {Digit}6 |

| Customer-Name | = | 1 {Character}39 |

| Dollar-Amount | = | "$" + 1 {Digit}9 |
| Credit-Ref-No. | = | "R" + 5 {Digit}5 |

A valid test input that might be included in the NPTS for Credit-Trans might be something like:

"A555555" + "Sebastian Dangerfield" + "$250"

All the components of a test input data flow are literals.

In set terminology, the number of entries in a given set, in an NPTS for instance, is referred to as the *order* of the set. What is the order of the NPTS for Credit-Trans? How many test input data flows will be required to make a complete normal path test set for this data flow?

In general, the order of an NPTS will depend on characteristics of the data elements. In our example, there are four data elements: Account-No., Customer-Name, Dollar-Amount, and Credit-Ref-No. For each data element, we will require the use of one or more test values to verify proper processing of that data element by the system. The set of test values selected for a given data element is called its test value set or TVS. A reasonable test value set for the item Dollar-Amount might be:

TVS(Dollar-Amount)   = $0, $1, $555555, $999999999

(I have left out the quotation marks since all the values are literals — I will continue to express unmixed sets of literals without quotation marks.) The order of this set is four, since there are four values.

In keeping with good test practice, I have chosen to include in my TVS each of the boundary values for the data element plus one mid-range value. A boundary value is one which is at some extreme of an allocated range. It is the shortest, the longest, the smallest, the largest, the whateverest possible value for that data element. (For a good discussion of boundary values, see Kernighan and Plauger, *Elements of Programming Style,* noted in your Bibliography.)

A TVS for each of the other data elements is listed below:

TVS(Account-No.)   =   A000000, A555555, A999999

TVS(Customer-Name)   =   A, Steven Dedalus, 39{Z}39

TVS(Credit-Ref-No.)   =   R00000, R55555, R99999

The four test value sets are of order 4, 3, 3, and 3, respectively. To determine the order of the composite, Credit-Trans, express its definition in terms of simple data flows, data flows with no internal operators other than +:

Credit-Trans   = | Account-No. + Customer-Name + Dollar-Amount |
| Account-No. + Customer-Name + Credit-Ref-No. |
| Account-No. + Dollar-Amount |
| Account-No. + Credit-Ref-No. |

The manipulation of DD expressions works just as manipulation of algebraic expressions, since the operators are commutative and associative.

The order of the NPTS for Credit-Trans is the sum of the orders of the component test sets for the four options. The order of each option is the product of the orders of the TVS's of the concatenated data elements:

$$
\text{Order of NPTS (Credit-Trans)} \quad = \quad
\begin{aligned}
& 3 \times 3 \times 4 \\
& + 3 \times 3 \times 3 \\
& + 3 \times 4 \\
& + 3 \times 3
\end{aligned}
$$

$$
= \quad 84
$$

I have treated this as a problem in combinations and permutations. The order of the NPTS is the number of permutations of the test values, taken in the four combinations represented by the four options.

The actual test input data flows of the NPTS are derived in the same fashion by combining and permuting the test values for the various data elements. I reproduce the first few test inputs of the NPTS and the last:

Test Input No. 1	=	A000000 + A + $0
Test Input No. 2	=	A000000 + A + $1
Test Input No. 3	=	A000000 + A + $55555

.
.
.

Test Input No. 84	=	A999999 + 39 (Z)39 + R99999

There is one NPTS for each input data flow. The NPTS's are combined into a normal path test data base. Any or all of the inputs in the normal path test data base may be applied during acceptance testing.

In some cases, the size of this data base can become cumbersome. It may not be possible for the data base ever to be completed. But all parties to the development project must understand that the system is expected to be able to handle each input in the data base, as well as any input that could ever be put into the data base through the process of selection of test values for each data element and permutation and combination of these values into test inputs.

26.2 Derivation of exception path tests

Next we require a set of inputs to test the system's capacity to deal with errors in input. I name this set of test inputs the exception path test set or EPTS. In order to derive the EPTS, we proceed as before by selecting values for the data elements. This time we select *invalid* values for each data element. The invalid test value set (ITVS) for Dollar-Amount might be:

ITVS (Dollar-Amount)   = null, $, $5555555555, 555, $abc, abc

These values are now varied into test inputs from the normal path test set in place of the item Dollar-Amount. The results are combined into an EPTS. There is one EPTS per *data element* present in any of the system inputs. The order of the EPTS is equal to the order of the ITVS for the data element. Note that the exception path tests each involve a single error. No attempt is made here to test multiple errors in the same input or combinations of errors in several inputs — that will come in the transient state tests.

In addition to the exception path tests that can be conducted on a per data element basis, there must be a set of tests to deal with format problems. (Certain errors of format obscure the divisions between data elements, so testing by data element would be useless.) In order to derive the format exception tests, we need to look at the *physical* composition of the data flow. The data flow Credit-Trans, as defined, could be entered in a fixed-field format, or key-word format, or prompted format, or many other physical formats. Each of these physical possibilities could be described in a Data Dictionary definition. The definition presented earlier for Credit-Trans would be the logical equivalent of all the physical variants.

Suppose we settle on a single physical format for multiple input types:

All-Trans-Format	=	Trans-Card + 1 {Field-Card} 25 + End-Card
Trans-Card	=	Trans-Name + {blank}
Field-Card	=	8 {Character} 8 *Field name* + 72 {Character} 72 *Field value*
End-Card	=	"END" + {blank}
Trans-Name	=	"CREDIT" "DEBIT" "INQUIRE" etc.

To test format error detection of our system, we need to prepare input sets that have invalid test values for each of these physical (format-dependent) data elements. We proceed, as above, to derive an invalid test value set for each data element and then vary them into test data flows from the normal path test data base. Care should be exercised to create tests for all the relevant boundary situations: null transaction card, no field cards, too many field cards, etc.

All the EPTS's and the format-dependent exception tests are combined into an exception path test data base.

26.3 Transient state tests

The tests derived thus far would be sufficient for a system that had no memory at all, one that treated each input in a fixed manner independent of what had happened before. Such tests will be useful in testing any system, but to cope with the various transient states that more complicated systems take on, it will be necessary to group the tests into sequences. The sequences of tests can then be arranged to force the system into each of the possible states and test its performance in that state.

To derive a complete set of test sequences, begin by listing all the memory states that pertain to each input data flow. Consider the states for the Credit-Trans example:

- No match on Account-No.

- No match on name

- Credit-limit surpassed

- No match on Credit-Reference-No.

- Account status is marked "HOLD"

- All tests pass

- All tests pass but new collateral required

There is one sequence for each state. Each sequence is a number of normal path test inputs that are to be entered successively to attain and exercise the target state. For state five, for example, the sequence might consist of several transactions on the same account: one to open the account, another to set status, and finally a valid Credit-Trans. If you have used the same mid-range Account-No. throughout your normal path test data base, you should be able to find the test input data flows there, rather than create more. To complete the test, you need to return the system to its initial state (remove the account), so that the next test is unaffected.

26.4 Performance tests

Performance tests check that the system measures up to the quantitative standards set for it in the Target Document. There is one test or set of tests for each performance restriction (throughput, response time, capacity, etc.) imposed upon the system. I have no great insight to offer you on performance testing. The requirements are so varied that it is impossible to say anything very useful about performance testing in general. I suggest that a summary of performance requirements and criteria for testing compliance should be published as part of the Structured Specification. This early publication of the test method may be a clearer statement of the true performance requirement than any other.

26.5 Special tests

It would be naive to think that any system, no matter how complex, could be tested completely by the derived test sets as I have explained them. Clearly some systems require additional tests to deal with their very particular natures. For such special tests I suggest this protocol: Since they cannot be derived from the Structured Specification, they must be placed explicitly in the Structured Specification before it is considered complete. This will maintain the standard that all the criteria for acceptance are contained in the Structured Specification.

26.6 Test packaging

Each test or test sequence should be packaged to include these parts:

- test procedure
- test data
- predicted test result
- procedure to return system to initial state

In the ideal situation, the package is in some form that allows testing to be managed by a computer rather than by people. The computer executes the test procedure, enters the test data, compares output to the prediction, reports on the comparison, and then returns the system to the initial state. Packaging tests in this fashion facilitates administration of large test sets and also allows for regression testing when changes are introduced.

27 HEURISTICS FOR ESTIMATING

A little joke on me to begin this section: It has taken me a lot longer than I estimated to get to this point. In spite of that, I shall carry on undaunted with my observations on estimating, just as if I had never missed an estimate myself. Even if I had a perfect record on estimating, anything I told you about it would be subject to this qualification: A technique or tool may be useful to the estimating process — it may even be necessary — but no single tool, and no set of tools, will ever be sufficient.

The difference between an approach that is sufficient to guarantee a certain result, and one that does not guarantee but often helps achieve that result, is the difference between an algorithm and a heuristic. An algorithm is a procedure that always provides a guaranteed solution. A heuristic is a cheap trick that *often* provides a good solution.

One example before we press on. Suppose you need to know how much time it will take you to build a brick wall according to a given specification. An algorithm to answer this question is to build the wall and time yourself; a heuristic would be to count the number of bricks and multiply by the net average time it has taken you to lay one brick in the past. For it to be useful input to this process, your observed bricklaying rate ought to be weighted to include mortar mixing, rest time, cleanup, and other factors that affect net productivity. The validity of the heuristic estimate, of course, depends on the applicability of your experience. Even though no two jobs are alike, use of adjusted productivity rates from the past usually gives a better quality estimate than any other method. This heuristic approach to estimating is one reason that bricklayers are better estimators than programmers.

There are no algorithms for estimating. The terms are mutually exclusive. It is possible, however, to develop some useful heuristics. These heuristics will sometimes let you down, but I suggest that there is no better approach to estimating.

The development of estimating heuristics will involve the following:

- the concept of an empirically derived estimate
- empirical productivity data

- estimating rules

27.1 The empirically derived estimate

An estimate is empirically derived to the extent that it makes use of past productivity information to predict future development. In order to derive empirical estimates for system projects, you need two things:

1. an early quantum indicator of the scope of development

2. empirical data that correlates observed performance with the same early quantum indicator on past projects

For the bricklayer, the early quantum indicator was the number of bricks and the empirical correlation was his past rate of laying bricks.

All that is very fine. If I tell you that yours is a 37-persnickety project and that past development in your organization has cost 11 work-months (plus or minus 6 percent) per persnickety, you and your calculator will quickly derive an empirical estimate. But in the absence of such a broad hint, it's not so easy. What are you to use as the early quantum indicator of scope, and where are you going to find the empirical correlation data?

I propose that you use the number of functional primitives in your Structured Specification as the early quantum indicator. In my own experience, this factor has had a strong correlation (plus or minus 20 percent) to the size of the final system measured in net lines of code delivered. And it has also had a strong correlation to the manpower needed to deliver the final system.

As for the correlation data, I can only suggest that you begin to collect it now. For your pilot projects in Structured Analysis you have to keep accurate, meaningful records on the use of development manpower. These records are then correlated to the number of functional primitives in order to calculate factors for predicting future development, factors such as

- total net manpower required to deliver the system, expressed in hours per primitive
- net manpower required per phase
- test resources used per primitive
- librarian time required per primitive
- number of specification changes per primitive
- overhead cost per primitive

My answer may not be altogether satisfactory, but surely you didn't expect it to be. I have not given you an estimating formula, but rather a formula for deriving an estimating formula. Also, my approach suffers from some of the same problems that have plagued past efforts to estimate based on net lines of code: productivity variations, language variations, variations of system complexity.

I advocate this approach, in spite of its inherent problems, because its problems are surmountable. Through the use of empirical derivation, we can hope to change estimating from a black art to a learnable skill. Use of the count of functional primitives gives us a much earlier indication of project scope than we ever had before. As for the variations in productivity and performance, they do exist — they are not a justification for ignoring empirical data, however, but rather an indication that we must collect as much of it as possible. And we must allocate more energy and care to determining the factors that make it vary.

27.2 Empirical productivity data

For years, our industry has staunchly refused to collect data about development performance. The arguments against it were these:

- Performance varies so widely that such data would be useless.

- Management is likely to use the collected data against poor performers. Worse yet, it is liable to misuse the data, since the reasons for performance variations will not be understood by upper levels.

The first of these points is defeatist in nature; it implies that we will never understand why some developments are more difficult than others, and we will never be able to predict the level of difficulty. If you feel that way, I urge you to look into Maurice Halstead's new book, *Elements of Software Science,* referenced in the Bibliography. Halstead makes a rather convincing case that it is possible to predict complexity of software; it is just more involved than you might have thought.

The use and misuse of performance data by management is a political problem and it deserves a political solution. I know of companies that have collected the information with a hands-off policy: Managers understood that information on individual productivity would not be made available to them.

How can you collect performance data so that it will be useful for predicting future development? The answer lies in keeping complete records of project proceedings, very much the way ships' officers keep logs. You must endeavor to collect data not only on those aspects that are unquestionably relevant, but also on those aspects that *might* be relevant.

When you have collected statistics on enough projects, you will be able to produce empirical estimating formulas that take into account variation in performance due to such factors as

- project duration
- team size
- overtime expenditures
- machine time for testing
- turnaround
- on-line test aids
- interface complexity
- coding language

You will also be able to establish tolerances for deviation from the predicted result.

27.3 Estimating rules

Some of your estimates are subject to empirical derivation, but others are not. What can you do if you have no early quantum indicator, or if there is no past performance to which you can compare it? In such a case, you are reduced to making an estimate in the form of an educated guess. But even this process is subject to some rules.

My first few estimating rules have to do with a rather curious but widely observed phenomenon: Most estimating failures are not merely failures to estimate accurately, but failures to estimate at all. Suppose, as an example, that your boss rushes in with this request:

> "Quick, give me your estimate for how long it's going to take you
> working full time without help to do suchandsuch between now and
> this time next year."

If you respond "one year," your boss will conclude that you are a splendid estimator and depart, well pleased. But you have not estimated anything. You have only regurgitated.

Rule 1: Estimating is different from regurgitating.

We are often called upon to regurgitate in the guise of estimating. Sometimes our superiors are subtler in letting us know what the "right answer" is. They might use an arched eyebrow or a soto-voce hint. But if there *is* a right answer and everyone knows it, then you are not even being asked to estimate, but to regurgitate.

Suppose you refuse to regurgitate like a good fellow, and respond by saying, "It can't be done by this time next year. I need 14 months." You are probably still not estimating. You may be negotiating.

Rule 2: Estimating is different from negotiating.

The difference between the two is the effect that the known right answer has on the process.

If you negotiate instead of estimate, you have started off on the wrong foot. Your sin, however, is not nearly as serious as if you first estimate and *then* negotiate. When you come up with your best estimate of an unknown, it makes no sense at all to let someone, namely your boss, make a counter offer.

Rule 3: Estimations are not subject to bargaining.

It is well to cultivate an air of stunned disbelief to greet any attempt at bargaining. At the very least, you are obliged to tell whoever makes the counter offer that the estimate is now his.

Suppose your boss asks you to apply your estimating skills to the scheduling process. "Make us up a schedule for three people to complete the Python-X project in one year." If you now lay out a schedule showing duration of analysis, design, and coding — you are still not estimating. You are dividing.

Rule 4: Estimating is different from dividing a fixed duration into component parts.

Let's say that you have produced a schedule and started work on the project. Due to bad luck, one of the early phases finishes late. If you now "estimate" that a subsequent phase is bound to finish early enough to make up the difference, you have once again not estimated. This time you have indulged in a fit of hysterical optimism.

Rule 5: A slip in one project phase implies a proportionate slip in all subsequent phases.

Rules 1 through 5 all deal with a failure to estimate due to the influence of a known "right answer." The very presence of such an answer inhibits the estimating process. You may not be able to control how estimating requirements are presented to you, but if you solicit estimates from your subordinates, keep this in mind:

Rule 6: If you want a meaningful estimate from someone, don't tell him "the answer."

If you do, he will regurgitate, negotiate, or divide.

So much for what estimating is not. We need a good working definition of the concept of an estimate. I offer you this one: *An estimate is a projection based on probabilistic assessment.* When you estimate that a given project can be completed in so many months, you are making a probabilistic statement. You are saying, at the very least, that the probability of finishing in that time is not zero. From my long and melancholy acquaintance with the estimating process, I conclude that most people do not mean to imply anything more by their estimates than the existence of a non-zero probability. In the absence of any better definition, they understand an estimate to be the most optimistic imaginable result that is not demonstrably impossible. That kind of estimate is a disaster for planning purposes.

> *Rule 7:* A useful planning estimate is a projection that is as likely to
> be too pessimistic as it is to be too optimistic.

In order to determine whether an estimate qualifies according to this rule, you have to compare the likelihood of it being off in either direction by some substantial percentage (say 50 percent). If you conclude that a project is as likely to finish 50 percent early as it is like to finish 50 percent late, then the estimated project duration qualifies.

Unfortunately, most people are not very good with probabilities and are therefore unlikely ever to come up with useful planning estimates according to Rule 7. For such people, the only workable approach to estimating is to let them calculate estimates based on the most optimistic assumption, and then to apply some factor to calculate the useful planning estimate.

> *Rule 8:* The ratio between the most optimistic estimate and a useful
> planning estimate is fairly uniform for any individual.

So it makes sense to calculate this factor (by comparing past estimates to past results) and to apply it for each individual who makes estimates for you.

For this approach to be useful, you need to inhibit feedback to the estimator. Tell him you want his most optimistic estimate, and don't complain when it's wrong.

Who should prepare estimates? In a series of estimating exercises at Yourdon inc., we found that the average person responsible for development was a worse than average estimator when it came to predicting anything about the development. In fact, for 49 test projects, the project manager's estimate was worse than the average of the other participants in all but one project! By appointing someone responsible for an effort, you destroy his ability to estimate effectively.

> *Rule 9:* Estimate by committee. Solicit estimates from everyone
> who understands the development. Throw away the estimate of the
> person directly responsible for the effort. Average the others.

My final word on estimating is this: Estimates deal with the unknown, and the unknown has a perverse way of subjecting poor developers to all kinds of rude shocks. I know of only one thing that keeps these rude shocks to a minimum, and I shall take this opportunity to pass it on to you: Good Luck!

GLOSSARY

algorithm   a procedure that leads to a guaranteed result

alias   a synonym for the Data Dictionary name of any data flow, data element, or file

analysis   the study of a business area prior to implementing a new set of (possibly automated) procedures

balancing   the relationship that exists between parent and child diagrams in a properly leveled Data Flow Diagram set; specifically the equivalence of input and output data flows portrayed at a given bubble on the parent diagram and the net input and output data flows on the associated child diagram

codification   representation of a subject matter in some predetermined rigorous format

cohesion   measure of the strength of association of the elements within a module

connection   reference from one module to a data item or entry point defined inside another module

Context Diagram   top-level diagram of a leveled DFD set; Data Flow Diagram that portrays all the net inputs and outputs of a system, but shows no decomposition

couple   information passed from one module to another along a normal connection, or information shared between two modules by means of pathological connection

coupling   measure of the interdependence of modules in a design structure; the amount of information shared between two modules

data base	data store that is accessed in more than one way, and that can be modified in format without affecting the programs that access it
Data Dictionary	set of definitions of data flows, data elements, files, data bases, and processes referred to in a leveled DFD set
Data Dictionary processor	program that affects a set of Data Dictionary procedures; specifically a program that allows definition control, and produces listings portraying definitions and relationships among definitions
data element	primitive data flow, one that is not decomposed into subordinate data flows
data flow	a pipeline along which information of known composition is passed
Data Flow Diagram (DFD)	a network of related functions showing all interfaces between components; a partitioning of a system and component parts
data store	repository of data; a time-delayed data flow; a file
Data Structure Diagram (DSD)	a graphic tool to portray relationships between data elements in a file structure
file	data store
flowgraph	a diagram that traces the stream of consciousness of person or machine as it carries out some policy
functional primitive	lowest-level component of a Data Flow Diagram; a process that is not further decomposed to a subsequent level
Functional Specification	classical product of analysis; description of a system to be implemented
heuristic	a procedure that often leads to an expected result, but makes no guarantee to do so
HIPO	Hierarchy plus Input-Process-Output, a technique for representing the modules of a system as a hierarchy and for documenting the insides of each module
human language	see natural language

incremental model	model of a portion of a system; model of a portion of a system as it is proposed in an associated change request; description of a proposed modification to a Structured Specification
information sink	net receiver of system information
leveled	portrayed in a hierarchical fashion such that the relationships among elements are presented as a tree structure
logical	implementation-independent; pertaining to the underlying policy rather than to any way of effecting that policy
mini-spec	Transform Description; statement of the policy governing transformation of input data flow(s) into output data flow(s) at a given functional primitive
model	representation of a system using DFD's, Data Dictionary, Data Structure Diagrams, etc.
natural language	language spoken by people, as opposed to a formal language, a language used by computers, or a metalanguage (a limited facility for rigorous description of a given logic)
normal connection	reference from one module to the name of another module; specifically a CALL
NPTS (normal path test set)	set of acceptance tests to exercise normal path processing (processing of input containing no errors)
orthogonal	property of a representational technique or descriptive method in which the functions of the various tools used do not overlap each other
pathological connections	reference from the inside of one module in a system structure to something defined inside another module other than that module's name
Petri Networks (PetriNets)	a network of related functions in a business operation in which people are portrayed as nodes, and documents as connections between nodes
physical	implementation-dependent

process	transformation of input data flow(s) into output data flow(s)
process description	mini-spec; statement of the policy governing transformation of input data flow(s) into output data flow(s) at a given functional primitive
SADT	Structured Analysis Design Technique, a proprietary data-flowing convention of SofTech Inc., Waltham, Mass.
schema	set of relationships among data elements in a complex file structure
specification increment	description of a proposed change of requirement in a format (Data Flow Diagrams, Data Dictionary, Structured English, Data Structure Diagrams, etc.) that facilitates integration into the Structured Specification; also Specification Increment Document (SID)
Structure Chart	graphic technique for portraying a hierarchy of modules and the relationships among them (specifically their connections and coupling)
structured	limited in such a way as to increase orthogonality; arranged in a top-down hierarchy
Structured Design	design technique that involves hierarchical partitioning of a modular structure in a top-down fashion, with emphasis on reduced coupling and strong cohesion
Structured English	a subset of the English language with limited syntax, limited vocabulary, and an indentation convention to call attention to logical blocking; a metalanguage for process specification
Structured Specification	end-product of Structured Analysis; a Target Document (description of a new system of automated and manual procedures) made up of Data Flow Diagrams, Data Dictionary, Structured English process descriptions, Data Structure Diagrams, and minimal overhead
subschema	portion of a schema; description of a private model of a file structure as conceived by a single user
system	connected set of procedures (automated procedures, manual procedures, or both)

system model

representation of a system using Data Flow Diagrams, Data Dictionary, Data Structure Diagrams, etc.

Target Document

the end-product of analysis; description of a system to be implemented — in order to be characterized a Target Document, the description should include *all* of the criteria for project success

test value sets (TVS)

set of selected data element test values to be used during acceptance testing

transaction analysis

a design strategy for original derivation of a modular structure from a Data Flow Diagram describing the policy; a design strategy that is applicable to portions of the Data Flow Diagram where there is parallel flow of similar data items by type

transform analysis

a design strategy for original derivation of a modular structure from a Data Flow Diagram describing the policy; a strategy that is applicable for the transform portions of a requirement, the portions that correspond to the shell of the structure chart (input legs, output legs, and location of the president module)

Transform Description

statement describing the logical policy that governs transformation of input data flow(s) into output data flow(s) at a given functional primitive

BIBLIOGRAPHY

Boehm, Barry W. "Software Engineering," *IEEE Transactions on Computers*, Vol. C-25, No. 12, December 1976.

Bohm, C., and G. Jacopini. "Flow Diagrams, Turing Machines, and Languages with Only Two Formation Rules," *Communications of the ACM*, May 1966, pp. 366-371.

Caine, Stephen H., and Kent E. Gordon. "PDL — A tool for software design." *National Computer Conference Proceedings*, Vol. 44, 1975, pp. 271-276.

Constantine, Larry L. (see Yourdon and Constantine).

Dahl, O.J., E.W. Dijkstra, and C.A.R. Hoare. *Structured Programming*. New York: Academic Press, 1972.

Flesch, Rudolf. *The Art of Plain Talk*. New York: Collier Books, 1946.

Dijkstra, E.W. (see Dahl).

Gane, Chris, and Trish Sarson. *Structured Systems Analysis: tools and techniques*. New York: Improved System Technologies Inc., 1977.

Gildersleeve, Thomas R. *Decision Tables and Their Practical Application in Data Processing*. Englewood Cliffs, N.J.: Prentice-Hall, Inc., 1970.

Halstead, Maurice H. *Elements of Software Science*. New York: Elsevier, 1977.

IEEE Transactions on Software Engineering, Vol. SE-3, No. 1, January 1977. (entire issue devoted to Structured Analysis)

Jackson, M.A. *Principles of Program Design*. New York: Academic Press, 1975.

Kernighan, Brian W., and P.J. Plauger. *The Elements of Programming Style*. New York: McGraw-Hill, 1974.

Miller, G.A. "The magical number seven, plus or minus two: Some limits on our capacity for processing information," *Psychological Review*, Vol. 63, March 1956, pp. 81-97.

Myers, Glenford J. *Reliable Software Through Composite Design*. New York: Petrocelli/Charter, 1975.

Orr, Kenneth T. *Structured Systems Development*. New York: YOURDON Press, 1977.

Stevens, W.P., G.J. Myers, and L.L. Constantine. "Structured design." *IBM Systems Journal*, 1974, Vol. 13, No. 2, pp. 115-139.

Teichroew, Daniel, and Ernest A. Hershey III. "PSL/PDA: A Computer-Aided Technique for Structured Documentation and Analysis of Information Processing Systems." *IEEE Transactions on Software Engineering*, Vol. SE-3, No. 1, January 1977.

Warnier, Jean Dominique. *Logical Construction of Programs*. 3rd ed. New York: Van Nostrand Reinhold Co., 1974.

Weinberg, Gerald M. *An Introduction to General Systems Thinking*. New York: John Wiley & Sons, 1975.

Yourdon, Ed. *How to Manage Structured Programming*. New York: YOURDON inc., 1976.

—. *Structured Walkthroughs*. New York: YOURDON inc., 1977.

—. *Techniques of Program Structure and Design*. Englewood Cliffs, N.J.: Prentice-Hall, Inc., 1975.

—, and Larry L. Constantine. *Structured Design*. 2nd ed. New York: YOURDON Press, 1978.

INDEX

667029